Pro Tools 11

Pro Tools 11
Music Production, Recording, Editing, and Mixing

Mike Collins

Focal Press
Taylor & Francis Group

NEW YORK AND LONDON

First published 2014
by Focal Press
70 Blanchard Road, Suite 402, Burlington, MA 01803

and by Focal Press
2 Park Square, Milton Park, Abingdon, Oxon OX14 4RN

Focal Press is an imprint of the Taylor & Francis Group, an informa business

Notices
Knowledge and best practice in this field are constantly changing. As new research and experience broaden our understanding, changes in research methods, professional practices, or medical treatment may become necessary.

Practitioners and researchers must always rely on their own experience and knowledge in evaluating and using any information, methods, compounds, or experiments described herein. In using such information or methods they should be mindful of their own safety and the safety of others, including parties for whom they have a professional responsibility.

Product or corporate names may be trademarks or registered trademarks, and are used only for identification and explanation without intent to infringe.

Library of Congress Cataloging in Publication Data
Collins, Mike, 1949-
   Pro tools 11 : music production, recording, editing, and mixing / Mike Collins.
      pages cm
   ISBN 978-0-415-81459-1 (pbk.)—ISBN 978-0-203-06641-6 (ebook)
1. Pro Tools. 2. Digital audio editors. I. Title.
   ML74.4.P76C67 2013
   781.3'4536—dc23   2013035185

ISBN: 978-0-415-81459-1 (pbk)
ISBN: 978-0-203-06641-6 (ebk)

Typeset in Myriad Pro
Project Managed and Typeset by: diacriTech
Printed in Canada

MIX
Paper from
responsible sources
FSC® C011825
www.fsc.org

Bound to Create

You are a creator.

Whatever your form of expression — photography, filmmaking, animation, games, audio, media communication, web design, or theatre — you simply want to create without limitation. Bound by nothing except your own creativity and determination.

Focal Press can help.

For over 75 years Focal has published books that support your creative goals. Our founder, Andor Kraszna-Krausz, established Focal in 1938 so you could have access to leading-edge expert knowledge, techniques, and tools that allow you to create without constraint. We strive to create exceptional, engaging, and practical content that helps you master your passion.

Focal Press and you.

Bound to create.

We'd love to hear how we've helped you create. Share your experience:
www.focalpress.com/boundtocreate

Table of Contents

About the Author

Mike Collins is a studio musician, recording engineer, and producer who has worked with all the major audio and music software applications on professional music recording, TV, and film scoring sessions since 1988.

During that time, Mike has regularly reviewed music and audio software and hardware for magazines including Future Music, Computer Music, PRS for Music's M magazine, Macworld, MacUser, Personal Computer World, Sound On Sound, AudioMedia, Studio Sound, Electronic Musician, EQ, MIX, and others. Mike also writes industry news and technical reports for Pro Sound News Europe.

Mike has been writing for Focal Press since 2000. His first book, 'Pro Tools 5.1 for Music Production', was published in December of that year. A second book, 'A Professional Guide to Audio Plug-ins and Virtual Instruments' was published in May 2003. 'Choosing & Using Audio & Music Software', Mike's third title for Focal Press was published in 2004 along with 'Pro Tools for Music Production Second Edition'. Mike's fifth book, 'Pro Tools LE & M-Powered', was published in the summer of 2006.

'Pro Tools 8' was published in the summer of 2009 and 'Pro Tools 9' followed this in the summer of 2011.

In 2010, Mike was invited to join a team of audio transfer engineers at Iron Mountain's Xepa Digital Studios in Slough, near London, to help to transfer a large part of Universal Music's back catalogue of classic popular recordings from analogue and digital tape copies of the archive stored in the United Kingdom to WAV files for archiving at Iron Mountain's secure underground facility in the United States. Using an Ampex ATR100 tape machine and Sonic Studio digital audio equipment, Mike personally transferred from ¼" tape much of the Chess catalogue along with many albums by Cat Stevens, Joe Jackson, Louis Armstrong, Quincy Jones, Barry White, Tricia Yearwood, and lots of other famous bands and artists from Universal's library of 'hit' recordings from the past 60 years.

Throughout 2011, Mike regularly performed 'live' as a duo with vocalist Aurora Colson at venues in and around London – see www.michaelandaurora.co.uk for more info. In the final quarter of 2011, Mike set up Rude Note Records – www.rudenoterecords.com – to release recordings he has produced, including Jim Mullen's solo jazz guitar album 'Thumbnail Sketches', together with three EPs and nine singles featuring Aurora Colson. Two collaborative albums featuring Mike with Jim Mullen, 'Blues, Jazz, & Beyond' and 'Pop, Rock, & Gospel', were released in December 2011 and February 2012, respectively. In April 2012, the label released Mike's remix of Vivienne McKone's song 'Everything is Gonna Be Alright', along with a 'cover' version of the Loose Ends hit, 'Hangin' On A String', featuring vocalist Joanna Kay. These recordings are all available from the Rude Note Records website via Spotify, iTunes, Amazon, and various other online retailers.

Between May 2012 and May 2013, Mike worked on a new album with singer/songwriter David 'DaPaul' Philips, recording and mixing the album, playing guitars on various tracks, and co-producing most of the tracks with David. This Gospel-influenced soul album, 'Soulful Spirit', was released in the summer of 2013 and is available at CD Baby –http://www.cdbaby.com/cd/dapaul2

Also active as a Music Technology Consultant, Mike Collins often presents seminars and chairs discussion panels on Pro Tools, Music Production, Music Technology, Music Rights, and Copyrights.

Contact Details

The author may be contacted via email at mike@mikecollinsmusic.com. The author's website can be found at www.mikecollinsmusic.com and a professional profile is available at www.linkedin.com/in/mikecollinsmusic.

Acknowledgements

First of all, I would like to thank Anaïs Wheeler at Focal Press for commissioning this book. I am also grateful for the efforts of Carlin Reagan and all the staff at Focal Press who are involved with publishing and marketing my books.

Louise Wells at Red Lorry Yellow Lorry PR – http://www.rlyl.com – was enormously helpful throughout the time I was writing this book, liaising constantly with Avid to provide review products and information. Thanks also to Avid's Bobby Lombardi who arranged to provide me with the beta software to allow me early access to Pro Tools 11 and extra thanks to Beta Program Manager Greg Robles who assisted me throughout the beta test phase with technical advice to keep my systems working. Avid UK Solutions Specialist Simon Sherbourne, as usual, was extremely helpful, supplying much-needed clarifications of various technical points.

Special thanks to my old friend Paul de Benedictis who came through at the last moment with some information from Bob Muller at Dangerous Music – http://www.dangerousmusic.com – clarifying how the Dangerous 2-bus is used with Pro Tools. Colin McDowell provided lots of useful information together with his entire suite of McDSP AAX plug-ins – http://www.mcdsp.com. Matt Ward, President at Universal Audio, and Erik Hanson, Director of Marketing at Universal Audio – http://www.uaudio.com – kindly supplied a UAD-2 system with its extremely comprehensive suite of plug-ins for Pro Tools.

Jean-Paul 'Bluey' Maunick, Kirk Whalum, and Dario Marianelli all took time out from their busy schedules to review and comment on the original edition of this book prior to publication for which I am extremely grateful. And record producer John Leckie also pointed out a couple of important things that I had overlooked in the Mixing chapter – explaining some basic points about mixing consoles that he had been shown during his training at Abbey Road Studios that really helped to clarify everything for me.

To all my regular musical collaborators and partners, including Jim Mullen, Lyn Dobson, John McKenzie, Winston Blissett, José Joyette, Marc Parnell, Vivienne McKone, Rouhangeze Baichoo, and David Philips, my wholehearted thanks for the musical inspirations and motivations with which they have kept me sane and fulfilled while I have developed my ideas for this book.

To Ernest Ranglin, Jamaica's foremost guitarist, who has not only been a major source of musical inspiration but has also acted as a personal and musical mentor – providing much-welcomed encouragement and positive feedback since we met back in 2009 – I offer my greatest respects and appreciation – see http://www.ernestranglin.co.uk for more info.

I would also like to thank all the members of my family, most importantly my father Luke Collins, my mother Patricia Collins, my brothers Anthony and Gerard Collins, and all my close friends, in particular Sia Duma, Dario Marianelli, Barry Stoller, Anthony Washington, Clive Mellor, and Keith O'Connell for their continuing support throughout.

Mike Collins,
August 2013

In this chapter

CHAPTER 1

Pro Tools – The world's leading digital audio workstation

Figure 1.1
Pro Tools 11
Splash Screen.

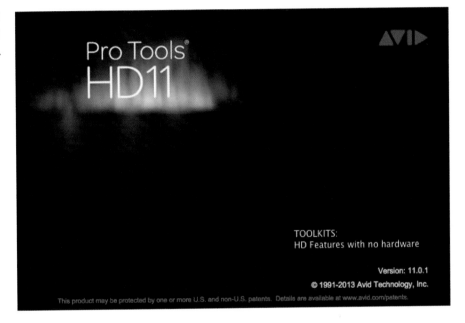

Introduction

Pro Tools digital audio workstation systems are used all over the world in professional recording situations: for 'live' recording at gigs and concerts, for film scoring, for audio-for-video and for audio post production – the list goes on and on…

This book will be particularly useful for people who are upgrading from previous versions of Pro Tools or 'cross-grading' from other digital audio workstations. It will also serve as a useful handbook to use alongside any Pro Tools system to clarify how things work and to provide useful operational tips and notes.

Although this book is not intended to be an entry-level text, I have explained things as simply as I can wherever possible. Mindful of the fact that some readers may not have English as their first language, I have tried to avoid colloquialisms and slang as much as possible and have done my best to write as clearly as possible. I hope that more experienced audio professionals will regard the sometimes-lengthy explanations of what they may consider to be basic points as useful reminders about how things work, which they can skip past easily enough if they wish.

There are several excellent books available that may be more suitable for beginners, which take a more step-by-step approach. And the Pro Tools Reference Guide and other Help documents are always available from the Pro Tools Help menu.

This book is about *music production* using Pro Tools. This is not to suggest for a moment that Pro Tools is not the leading system used for putting sound and music to picture or for mixing in surround. It is just a reflection of the fact that Pro Tools has truly come of age as a music production system – and the greater part of my experience is with music production, so I am naturally inclined to write about the things that I know most about. So the main focus of the book is firmly on using Pro Tools for music production and mixing in stereo – although surround formats have been with us for many years now. When did you ever hear of a Top 40 'hit' being released primarily as a 5.1 surround mix and making it because of this? I never heard of this happening – not one time! So this book will not digress into the fascinating world of surround sound – as interesting as this is to me and to many people I know.

Having read all the Pro Tools manuals and documentation in depth, along with most of the other books written about Pro Tools, I have noticed that some areas are not covered as thoroughly or as accurately as they could be, so I have made a point of addressing as many of these areas as possible. One example would be levels and metering, which are covered in plenty of detail in the chapter on Mixing.

You will not find every feature of Pro Tools covered in this book: it makes no attempt to be 'all things to all people'. What you will find is clear explanations of most of the things you definitely need to know about to record, edit, and mix audio and MIDI in Pro Tools in stereo. There is a substantial chapter on MIDI that covers all the technical stuff about Pro Tools that you need to get professional results, whether using hardware or software MIDI instruments.

Wherever possible, I have tried to explain topics that are covered in the Reference guide in a clearer or alternative way. Throughout the book you will find highlighted Tips and Notes. The tips are often taken from my personal experience, and the notes are often technical points taken from the Pro Tools Reference guide or other documentation that may otherwise be overlooked, so watch out for these.

Learning Pro Tools

The only way to learn Pro Tools is to use Pro Tools as often as possible in as many contexts as possible. Even if you own a Pro Tools HD system, I strongly recommend that you get yourself a laptop with Pro Tools software, so that you can take it wherever you go. This way, you can practice recording, editing, sequencing, and mixing Pro Tools sessions, or making music with virtual instruments and loops or whatever, while you are away from your main studio setup. Or you could get a compact desktop such as an iMac at home and practice using this with Pro Tools and an Mbox or other compact hardware.

If you are really determined to be the fastest Pro Tools operator around, you will have to 'eat, sleep and breathe' Pro Tools until you have learned as many of the keyboard commands as you can remember and get as much varied experience of recording as possible.

On the contrary, you may prefer to take a more relaxed approach – exploring the menus at a more leisurely pace rather than constantly typing commands on the keyboard. And Pro Tools has the advantage that it is simpler to get familiar with than most of its competitors – especially as it has just two main screens, the Mix and Edit windows.

Pro Tools 10

If you are upgrading from Pro Tools 9, you will be interested to learn about the changes that took place when Pro Tools 10 was introduced in 2011. For example, when Clip Gain was introduced, what was formerly the 'Region' menu underwent a name change and became the 'Clip' menu. For more details, read 'Appendix 13 – New for Pro Tools 10' on the companion website.

The Companion Website

The publisher hosts a companion website at:

www.taylorandfrancis.com/cw/collins

In addition to information about the book, here you will also find a set of useful Appendices:

Appendix 1 – About Summing Mixers

Appendix 2 – The Trouble with Dongles

Appendix 3 – Avid Eucon Support

Appendix 4 – Pro Tools LE Hardware

Appendix 5 – Pro Tools HD Hardware

Appendix 6 – The Eleven Rack & The Mbox Family

Appendix 7 – Pro Tools HD New Hardware 2010

Appendix 8 – Pro Tools 8 New Features

Appendix 9 – Pro Tools 9 Laptop System

Appendix 10 – Universal Audio Powered Plug-ins

Appendix 11 – New for Pro Tools 9

Appendix 12 – McDSP Plug-ins Roundup

Appendix 13 – New for Pro Tools 10

Pro Tools 10 | Pro Tools 11 Co-installation

Although Pro Tools 11 has been available since the summer of 2013, a number of users may still need to work with Pro Tools 10 for some projects so that they can continue to use any 32-bit TDM and RTAS plug-ins that are not available in the 64-bit AAX format that Pro Tools 11 requires.

Having both versions of the software installed on the same computer was not possible with older versions of Pro Tools. However, because version 11 uses the Avid Audio Engine while Pro Tools 10 uses the Digidesign Audio Engine (DAE) to interface the software with audio hardware, it is possible to install both versions of the software onto the same hard disk partition without any conflict.

For this to work successfully, you do need to be using version 10.3.5 (or higher) of the older Pro Tools software. With both software versions installed, it is only possible to run one of these at any given time – but it only takes a short time to quit one of these and then launch the other.

And why might you wish to do this? The main reason would be so that you can access older TDM and RTAS plug-ins that are not available in AAX format. Also, PT 11 does not use the Sound Designer II file format, so you would need to use PT 10 (or older) software to open and convert these files to WAV format for use with PT 11. And PT 11 will not open Pro Tools 4.x sessions or lower, so, if you need to access these older formats, this is another reason for keeping the Pro Tools 10 software installed.

You may also have computers with older operating systems that you still wish to use, such as a laptop or an iMac or an older Mac Pro that have Mac OS X 10.6.8 (Snow Leopard) or Mac OS X Lion installed, or a PC with Windows 7 Home Premium or Professional. Pro Tools 10 will run on these older operating systems, but Pro Tools 11 will only run on Mac OS X Mountain Lion (10.8) or Windows 7 or 8 Home Premium, Professional, or Ultimate (64-bit).

TIP
Pro Tools 11 and Pro Tools 10 have the same application icon and will look the same in the Dock, so you need to adopt some sort of strategy to deal with this situation. I simply rename one application as 'Pro Tools 11' and the other as 'Pro Tools 10' – or maybe just rename one of these so that it is clear which is which. Then, when you roll the mouse cursor over these icons in the Dock, the file name of the Application will appear as you hover over each icon – see Figure 1.2 screenshot. Another option is to place the icons in a different position in the Dock and simply remember which you placed in which position – such as the Pro Tools 11 icon to the right of the Pro Tools 10 icon.

Figure 1.2
Part of the Dock,
showing the mouse
cursor rolling over the
Pro Tools 11 icon.

> **NOTE**
>
> Pro Tools 11 sessions that use more than 4 gigabytes of RAM will cause Pro Tools 10 or lower to quit unexpectedly. The fix for this is to reduce the amount of session RAM usage to less than 4 gigabytes while working in Pro Tools 11 by removing virtual instruments that take up system memory and then saving this version to disk, safe in the knowledge that it will open successfully in Pro Tools 10.

Pro Tools 11 versus Pro Tools HD 11

Pro Tools 11 gives you 96/48/24 voices of simultaneous audio playback at 48/96/192 kHz, whereas Pro Tools HD 11 gives you 256/128/64 voices of simultaneous audio playback at 48/96/192 kHz per audio card.

PT 11 can record up to 32 tracks at once, whereas PT HD 11 can record up to 256 tracks at once (per card with HDX systems) – depending on the capabilities of your hardware.

Pro Tools 11 now supports up to 128 Instrument tracks, whereas Pro Tools HD 11 supports up to 256 Instrument tracks. Bear in mind that these are theoretical maximum numbers: any real-world system will be constrained by the demands placed on the computer you are using and by the particular virtual instruments that you are using.

Both can use up to 512 MIDI tracks and both can use up to 256 busses, but PT 11 only offers up to 160 Aux tracks, whereas PT HD 11 can use up to 512 Aux tracks.

The video features in PT 11 are quite good, with one HD video track available to play and edit, but the video features in PT HD 11 are much better, with up to 64 tracks of video available. Perhaps the best new feature is that you can add, play, and edit Avid DNxHD format files and other popular industry-standard HD video formats directly in the timeline – without having to transcode these first: a time-consuming step that was necessary with previous versions of Pro Tools.

Both Pro Tools 11 and Pro Tools HD 11 use the AAX plug-ins format, but systems that incorporate one or more Pro Tools|HDX cards can also use the AAX DSP format that runs on the processors on the HDX cards.

Features that are only available in Pro Tools HD 11 include Surround Mixing with up to 7.1 channels; Advanced metering with gain reduction metering; Advanced video editing with multiple video tracks and playlists; Track Punch/Destructive Punch; Input Monitoring; VCA mixing; Solo bus AFL/PFL; Advanced audio editing capabilities such as Continuous Scrolling, Scrub Trim Tools, Replace Clip command, field recorder workflows; and Advanced automation features such as Punch, Capture, Write on Stop, Write to All Enable, Automatch and Preview. Satellite link is also included, which allows you to control multiple Pro Tools|HD systems.

Complete Production Toolkit

Available for Pro Tools 10 and earlier versions, the Complete Production Toolkit expands the software to include many of the same features found in Pro Tools HD software. So, you get up to 256 simultaneous audio playback tracks (up from 96), 768 total audio tracks (up from 256), 128 instrument tracks (up from 64), and 64 video tracks (up from one). With the extended disk cache, you can have Pro Tools load your entire session into RAM to boost storage performance, system responsiveness, and reliability. For complex sessions with lots of automation, you can create VCA groups to get another layer of gain control over groups that already have automation. And you can add movement to your mixes beyond the standard stereo field – using up to 7.1 surround sound.

You also get TrackPunch/DestructivePunch; Input monitoring; Solo bus AFL/PFL; Advanced audio editing (Continuous Scrolling, Scrub Trim tool, Replace Clip command, field recorder workflows, and more); Advanced automation (Punch, Capture, Write on stop, Write to all enabled, Automatch, Preview, and more); and Advanced video editing (multiple video tracks, multiple video playlists, and video editing tools).

> **NOTE**
> The Complete Production Toolkit is not offered as an option with Pro Tools 11 software, so if you need these more advanced features, your only option is to buy Pro Tools HD 11.

Normally, Pro Tools HD 11 can only be purchased together with Avid hardware such as the Pro Tools/HDX or Pro Tools Native cards.

However, owners of PT 9 or PT 10 and the Complete Production Toolkit are allowed to upgrade to Pro Tools HD 11 software (which works with most hardware), without being required to purchase Avid hardware. These upgrades are priced at $999 and $599, respectively.

PT 11 Hardware Support

Pro Tools 11 will work with the following Avid systems and interfaces: Pro Tools|HDX; Pro Tools|HD Native; all Pro Tools HD Series interfaces (HD I/O, HD Omni, HD MADI); Mbox family (second and third generation); Eleven Rack; Fast Track Solo and Duo; Mojo DX; Nitris DX; VENUE systems with an FWx or HDx option card; and VENUE systems with a built-in Pro Tools interface.

Pro Tools HD systems no longer officially support the blue range of interfaces, including the 96 I/O, 96i I/O, 193 I/O, Sync HD, MIDI I/O, and PRE I/O – but these should still work.

The 003 and Mbox 2 families will be officially supported for Pro Tools 11. The 002 family, although 'Not Officially Supported', should continue to work. The Control 24 and Command 8 control surfaces are also 'Not Officially Supported', but will still work, while the HD Accel TDM PCIe and PCI card and the Pro Control are now 'Not Compatible' and will not work with PT 11.

You may also need a new iLok if you are still using an original iLok to authorize your Pro Tools software: Pro Tools 11 requires the license (or bundle that includes a Pro Tools 11 license) to be on a Second Generation iLok (iLok 2).

To get the latest information about hardware compatibility, go to Avid's support site at *www.avid.com/us/support* and type 'Pro Tools 11 compatibility' into the Search field to bring up a list of articles about System Requirements, Approved Audio Interfaces and Peripherals, Qualified Windows and Apple Computers, and so forth.

Summary

Pro Tools 11 offers improved workflows with much faster operation and enhanced performance overall compared with previous versions. In my view, Pro Tools HD11, especially when paired with one or more HDX cards, offers the best performance from a digital audio workstation that money can buy.

The changes to the Pro Tools software user interface going from version 8 through versions 9 and 10 to the current version 11 are relatively minimal. The biggest change was probably the renaming of regions as clips. This stability of the user interface design is a reflection of the reality that this design is working extremely well and an acknowledgement that it has been accepted by large numbers of users in preference to alternative software from rival companies. Avid's professional hardware systems are also used by most of the leading professional music-recording and film-scoring studios for the very good reasons that they provide the largest numbers of inputs and outputs, highest track counts, and high-quality interface options, together with on-board processing that delivers consistent results to the highest standards.

Proffering its reasons for discontinuing support for its older hardware, Avid says 'As a 64-bit application, Pro Tools 11 runs natively on 64-bit operating systems and hardware, but it cannot run on 32-bit operating systems. It also cannot support 32-bit hardware, such as Pro Tools|HD Accel hardware. This is a technical limitation due to the DSP chips on Pro Tools|HD hardware - Pro Tools|HD PCI and PCIe cards use a 32-bit architecture and cannot be modified to support 64-bit software. HDX and HD Native cards are 64-bit, as are many USB and FireWire interfaces (such as the third generation Mbox series of audio interfaces)'. So there you have it – straight from 'the horse's mouth'!

In this chapter

CHAPTER 2
Using Pro Tools

Introduction

Getting started with any software environment involves something of a learning curve and, if you are new to Pro Tools, you will find that this is no exception. Even if you have just upgraded from an earlier version, there may be a number of new features that you have not encountered previously. And if you are 'cross-grading' from another platform such as Apple Logic, MOTU Digital Performer, or Steinberg Cubase or Nuendo, you may be looking for features or trying to use keyboard commands that you are used to and wondering where these are in Pro Tools. So boot up your Pro Tools system and spend as much time as possible working 'hands-on' with the software until you know it inside out! Oh, and reading this book will help…

Getting Started

If you are using Pro Tools for the first time, you will need to spend some time familiarizing yourself with the user interface, or with its more recent features if you are upgrading. You will also probably find it helpful to configure the software to suit the ways you like to work. You should take some time out to practice using the system in a non-critical situation first (without impatient clients breathing down your neck) and take the trouble to learn at least a basic set of keyboard commands so that you don't have to use the mouse and menus all the time. You should also learn how to quickly find your way around whichever piece of music you are working on – zooming in and out and navigating along the timeline until you feel comfortable with all this.

Help

Avid provides excellent manuals for Pro Tools as Adobe Acrobat .pdf document files that you can access from the Help menu. When you open the Pro Tools Reference Manual, for example, the Acrobat application launches and the document file opens into a new window on your computer. You should resize this window and position it in a convenient place on your screen.

It can be very useful to keep the relevant manual open on your computer, but hidden until you need it. You can use the standard menu command or the keyboard Command-H on the Mac to hide the Acrobat application. When you want to reveal the manual again, you can always select it again from the Help menu or use the Show All menu command. One of the quickest ways to do this is to press and hold the Command key, then repeatedly press the Tab key until you see the application you want displayed in the middle of the screen. When the Tab has moved you along the list of open applications to Acrobat, let go of the Command key and this will be shown and brought to the front. Windows users will have their own preferred ways of doing these things.

Starting the Engines

As with any complex piece of machinery, there are various settings that you may need to adjust each time you want to use it. This is certainly the case with Pro Tools. I recommend that you always check the Playback Engine settings and the I/O Setup at the start of any new session. You may also need to make some changes to the Hardware Setup if you want to hook up additional hardware.

Avid Audio Engine

If you are using Avid hardware, Pro Tools 11 will use the Avid Audio Engine. This Audio Engine is a real-time operating system for digital audio recording, playback, and processing designed for use with Pro Tools and Avid audio hardware that is automatically installed on your system when you install Pro Tools. If you are using hardware made by another manufacturer, Pro Tools 11 will use Core Audio on the Mac or Audio Stream Input/Output (ASIO) on Windows.

Apple's Core Audio provides the audio connections between software applications, such as Pro Tools 11, Digital Performer, Cubase, Nuendo or Logic, and any audio hardware installed on a computer that uses Mac OS X.

Similarly, Steinberg's ASIO provides the audio connections between software applications such as Pro Tools 11, Digital Performer, Cubase or Nuendo, and audio hardware installed on a computer that uses Windows.

NOTE
Pro Tools software can only record up to 32 input channels of audio or play back up to 32 output channels of audio, when using audio interfaces with the Avid Audio Engine, or supported Core Audio (Mac), or ASIO (Windows) drivers. So if you need more channels for I/O (Input and Output), you will have to use Pro Tools HD software with Avid HDX or Native hardware.

Avid Video Engine

Avid's Video Engine works with QuickTime video, so you can use any video that you have available in this format. The Video Engine also works with a wide range of Avid HD and SD MXF video formats on Pro Tools video tracks without the need for transcoding these first, including Avid DNxHD. It also lets you monitor Avid HD and SD MXF, and QuickTime media using Avid Nitris DX, Mojo DX, and other Avid qualified third-party video interfaces.

> **NOTE**
> If you want to use video in your Pro Tools session, then you will need to tick the box in the Playback Engine dialog to enable the Video Engine – otherwise, the video track(s) in Pro Tools will not work.

The Playback Engine Dialog

In the Playback Engine dialog, Pro Tools provides a pop-up selector that allows you to choose the audio 'Playback Engine' for use with your audio interfaces. The available options will depend on which audio interfaces are connected and have compatible drivers installed.

Changing the Playback Engine can be useful if you have two or more audio interfaces connected to your computer with different routing configurations in your studio or if you want to prepare a session for use with a specific interface on a different system (e.g. you might want to prepare a session created on your Avid HDX system for use with the built-in audio on your Mac laptop).

On the Mac, for example, there will always be a 'Built-in' audio interface, and there may be another if you are using an Apple display monitor that has audio input and output capabilities. On Mac systems, you can also select the Pro Tools Aggregate I/O option, which lets you use a combination of built-in inputs and outputs at the same time – see Figure 2.1.

You can configure the I/O options for Pro Tools Aggregate I/O using the Audio Devices window from the Mac's Audio MIDI Setup utility software – choosing an appropriate combination of inputs and outputs – see Figure 2.2. These inputs and outputs will become available in the I/O Setup dialog if you open this and click on the Default buttons in the Insert, Input, and Output tabs.

The H/W (Hardware) Buffer Size pop-up selector in the Playback Engine dialog – see Figure 2.3 – lets you choose the size of the buffer used to handle host-processing tasks such as processing with host-based, or "Native" plug-ins.

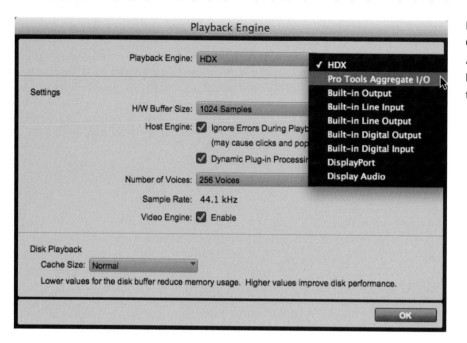

Figure 2.1
Choosing Pro Tools Aggregate I/O in the Playback Engine on the Mac.

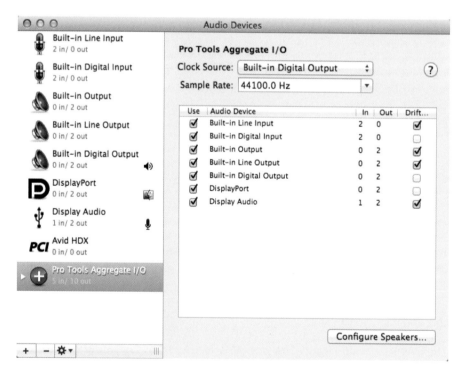

Figure 2.2
Configuring Pro Tools Aggregate I/O in the Audio Devices window.

Figure 2.3
Hardware Buffer Size
and Host Engine
settings.

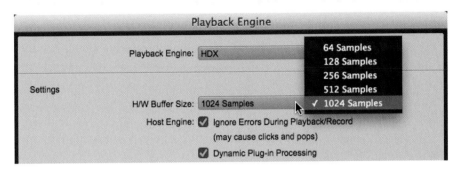

TIP
You can open the Audio MIDI Setup utility by choosing the MIDI Studio sub-menu item available from the MIDI menu item in the Setup menu in Pro Tools. This opens the MIDI window by default, but then you can open the Audio window from the Window menu in the Setup utility.

On all Pro Tools systems, lower settings reduce MIDI-to-audio latency – for example, the delay between playing notes on your MIDI keyboard and hearing the audio response from a virtual instrument. Lower settings can also improve screen response and the accuracy of plug-in and mute automation data.

If you are using lots of 'Native' plug-ins, you should choose a higher buffer size to allow for the greater amounts of audio processing required.

The Host Engine settings provide options for error suppression during playback and recording and the option to use dynamic plug-in processing.

There may be times when it makes sense to tick the box to Ignore Errors During Playback/Record. For example, when you are working with several instrument plug-ins and these are stretching the capabilities of your computer's CPU to the point where you are hearing clicks and pops in the audio. In this case, you may choose to work with reduced audio quality, accepting that there are clicks and pops in the audio while you are trying out arrangement ideas. Later, when you want to make sure that you are getting the highest possible audio quality, you can disable this option.

Ignoring errors requires at least 128 samples of additional buffering on some systems. Host-based Pro Tools systems have an option to Minimize Additional I/O Latency. Enabling this option restricts any additional latency due to ignoring errors during playback and recording to 128 samples. With this option

> **NOTE**
>
> On host-based Pro Tools systems, lower settings reduce all input-to-output monitoring latency on any record-armed tracks or Auxiliary Input tracks with 'live' inputs. On Avid HDX systems, lower settings reduce the monitoring latency that occurs on tracks that have one or more Native plug-ins. Lower settings can also improve the accuracy of MIDI track timing on systems without a MIDI interface that supports time stamping and on tracks using MIDI virtual instruments that do not support time stamping.

disabled, the buffer used for error suppression will be at least 128 samples or half the H/W Buffer Size – whichever is greater.

The Dynamic Plug-In Processing option maximizes plug-in counts by dynamically reallocating host-based processing resources as needed, so plug-ins only use CPU cycles when they are actually processing audio. Normally, you will want to make sure that this option is enabled.

> **TIP**
>
> If you are using a slower computer, you may want to disable the Minimize Additional I/O Latency option to avoid adverse performance.

> **NOTE**
>
> The Minimize Additional I/O Latency option is only available if the Ignore Errors During Playback/Record option is enabled and the Pro Tools system you are using requires additional buffering for error suppression, as is the case with the following: Mbox Pro and Mbox 2 Pro on Windows and the Mbox family devices, Digi 002 and 003 devices, Eleven Rack, and Pro Tools Aggregate I/O on the Mac.

Pro Tools HDX systems

On Pro Tools HDX systems, there is a Number of Voices setting in the Playback Engine dialog – see Figure 2.4 – that lets you choose the number of available voices and how those voices are allocated to DSPs in your system. Changing the number of voices affects DSP usage, the total number of voiceable tracks, and overall system performance. The maximum number of voices available will depend on the session sample rate and the number of Avid HDX cards in your system.

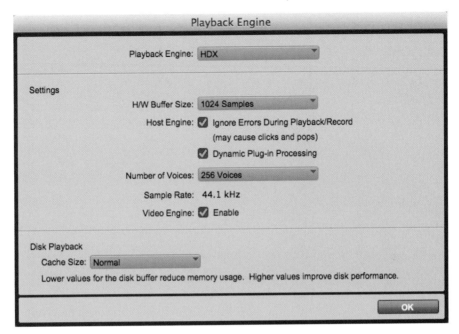

Figure 2.4
Playback Engine
dialog on an HDX
system.

▶ If you are using high-bandwidth PCIe cards (such as video capture cards) along with your Avid HDX cards, you should select the minimum number of voices.

▶ If your HD cards are in an expansion chassis, or when you are using other PCIe cards along with Avid HDX cards, select a medium number of voices.

▶ If your Avid HDX cards are the only PCIe cards in your computer, or when you are using an expansion chassis to run higher track counts (such as 64 tracks at 96 kHz) and you want more voices per DSP (such as 16 voices per DSP at 96 kHz), then you should select higher numbers of voices.

Pro Tools HD Software

The Playback Engine dialog in Pro Tools HD software has an additional section for Disk Playback Cache Size settings. Here, you can choose the amount of memory the Avid Audio Engine will use to pre-buffer audio for playback and recording using the Cache Size pop-up selector in the Playback Engine – see Figure 2.5.

Pro Tools HD loads audio files used in Pro Tools sessions into a RAM cache prior to playback, prioritizing files closest to the current play head location so that these are already in the cache when you start playback.

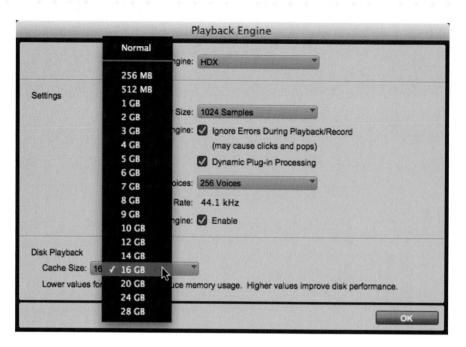

Figure 2.5

Disk Playback Cache
Sizes available
with 32 GB of RAM
installed on the
computer.

The Pro Tools HD software checks to see how much RAM you have installed on your computer and then subtracts 4 GB from this amount to work out the maximum amount of RAM that you could allocate to the Disk Cache.

For example, my computer has 32 GB of RAM installed, so the largest amount of RAM available for the Disk Cache is 28 GB – although, in practice, I normally choose a smaller amount unless I am working on a large session.

NOTE

The Sample Rate setting available in the Playback Engine when using HDX systems shows the default sample rate used when you create a new session. If a session is open, the current session sample rate is displayed, but cannot be changed. If no session is open, you can set the default sample rate for new sessions.

It is usually best to leave the Cache Size on the default setting of Normal, which is the optimum Cache Size. However, if you do select a fixed Cache Size in the Playback Engine dialog, a Disk Cache meter will appear in the Activity meters section of the System Usage window – see Figure 2.6.

The Activity meters section has Disk Cache and Memory meters that you can use to help you to decide whether to assign more or less RAM to the Disk Cache for the current session.

Figure 2.6
System Usage Window
Activity meters showing
the amount of Disk
Cache memory in use.

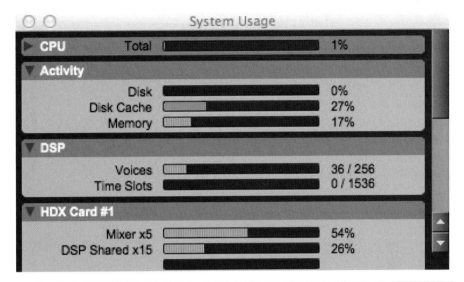

NOTE

The Disk Cache meter indicates the amount of allocated memory being used to cache audio in the timeline (up to the amount of RAM selected for the Disk Cache setting in the Playback Engine). An additional peak-meter style indicator is included to show how much of the allocated RAM is needed for caching the session. The Memory meter displays how much of the installed physical RAM in the system is being used by Pro Tools. This includes RAM used by the audio engine, the video engine (if enabled), plug-ins, and the disk cache. It does not show any RAM usage in your computer system other than by Pro Tools.

TIP

If the Memory meter approaches 100%, install more physical RAM or lower the Disk Cache setting in the Playback Engine dialog.

NOTE

Avid HDX systems provide additional meters below the Activity meter for Voices, Time Slots, and DSP:

The Voices meter displays the total number of voices that can be allocated and the number of voices currently allocated. This includes all voices

(Continued)

whether they are allocated explicitly or dynamically, as well as any voices used for routing host-based processing.

The Time Slots meter displays the total number of DSP Time Slots available and the number of DSP Time Slots currently used.

The DSP meters display how much of each DSP chip on each Avid HDX card is currently being used for mixer configurations and DSP-based plug-ins.

Hardware Setup

The Hardware Setup dialog (see Figure 2.7) lets you make various settings for your hardware interface or interfaces. On most systems, you will make these settings when you install your Pro Tools system and then leave them alone from that point onwards.

If you are using a third-generation Mbox family audio interface or a third-party Core Audio (Mac) or ASIO (Windows) compatible audio interface, a Launch Control Panel button is provided to launch the control panel for your particular audio interface to configure its settings.

If you are using an Mbox family or a third-party audio interface, there will be either a Launch Control Panel button or a Launch Setup App button (depending on which audio interface you are using) in the Hardware Setup dialog.

When you open this control panel or setup application, you can change the mixer, output, and hardware settings, including the sample rate, hardware buffer size, and sync source.

Figure 2.7 Hardware Setup with a Focusrite Saffire third-party audio interface.

Avid HDX and HD Native systems can have HD I/O, HD OMNI, or HD MADI audio interfaces connected to HDX cards or HD Native hardware. Large recording studios will be using some combination of HD I/O audio interfaces, each of which supports 16 channels of simultaneous I/O and multiple I/O formats (such as analog, AES/EBU, ADAT Optical, S/PDIF, and TDIF), or HD MADI interfaces, which support up to 64 channels of MADI I/O. Smaller studios are more likely to be using the HD OMNI interface, which supports up to eight channels of simultaneous I/O and multiple I/O formats (such as analog, AES/EBU, ADAT Optical, and S/PDIF). The HD I/O and HD OMNI can also have additional interfaces attached using the Expansion port on each interface.

NOTE

If you have multiple audio interfaces of the same type connected to your system, make sure that you choose the appropriate interface in the Peripherals list when you define its inputs and outputs in the Hardware Setup.

TIP

You can set the sample rate when creating a new Pro Tools session by selecting a different sample rate in the New Session dialog.

NOTE

On Mac systems using Core Audio, you can select Pro Tools Aggregate I/O as the Current Engine to use the built-in audio inputs and outputs on your Mac computer. You can configure the Pro Tools Aggregate I/O setting in the Mac Audio Setup, which can be accessed from the Pro Tools Hardware Setup dialog. The Pro Tools Aggregate I/O device is intended for use only with the built-in audio on your Mac computer. For best performance, use the default settings.

If you are using an Avid HD audio interface, the Hardware Setup dialog lets you configure the signal routing, digital I/O format, default sample rate, clock source, and other hardware-based settings for each HD peripheral connected to your system.

NOTE

You can also configure similar settings in the Hardware Setup dialog if you are using Avid 003 or 002 family interfaces.

For example, using an HD Omni interface, the Main page of the Hardware Setup dialog is where you define which physical inputs and outputs on your audio interface are routed to available inputs and outputs in Pro Tools. The numbers at the left of the central section (1–2 through to 7–8) are the numbers of the pairs of inputs and outputs that you will see in the Pro Tools software. The columns marked Input and Output have pop-up selectors that let you select which physical inputs or output will be routed to the Pro Tools inputs and outputs represented by the numbers to the left. In the example shown, in Figure 2.8, the pop-up selector has been used to select the S/P DIF input to be routed to Pro Tools input pair 5-6.

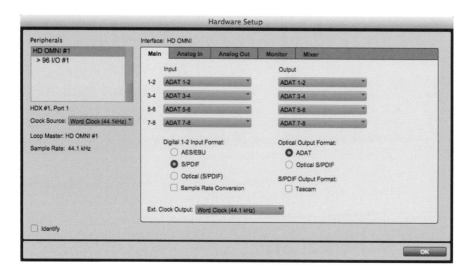

Figure 2.8
Hardware Setup for HD Omni Interface.

In the area of the Hardware Setup Window nearest to the left-hand side of this window, you can select the Clock Source for the system using the Clock Source pop-up selector. You can also change the Clock Source in the Session Setup window, in which case the changes will be reflected here. When you are recording analog audio directly into Pro Tools, you will normally choose the Internal clock source. If you are transferring material into Pro Tools from an external digital device, you will need to synchronize to that digital device – or to a house clock signal that you use to synchronize all your audio recording devices. External options can include AES/EBU, S/PDIF, Optical, TDIF, ADAT, and Word Clock.

The Sample Rate setting is also displayed at the left of the Hardware Setup window. When the session is open, this cannot be changed, but you can use this to change the default sample rate when you create a new session.

NOTE

With Avid HDX and HD Native hardware, you can change the default Sample Rate in the Hardware Setup or in the Playback Engine. On other Pro Tools systems, you can only change the default sample rate in the Hardware Setup or using the control panel for third-party audio interfaces. Be aware that you can always change the sample rate when creating a new Pro Tools session by selecting a different sample rate in the New Session dialog.

Additional pages are available in the Hardware Setup dialog and you can switch through the different pages either by clicking on the tabs marked Main, Analog In, Analog Out, Monitor, and Mixer or by holding down the Command key on your computer keyboard and pressing the Left or Right Arrow keys if you are using a Mac computer. Hold the Control key and press the Left or Right Arrow keys if you are using a Windows computer.

If you are working with the HD OMNI, Avid recommends that you configure the Monitor page first – see Figure 2.9.

Figure 2.9
HD OMNI Hardware
Setup, Monitor page.

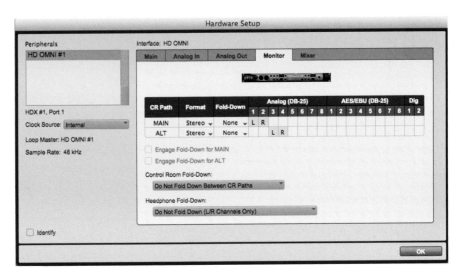

The Analog In page lets you set the operating levels for the inputs – see Figure 2.10.

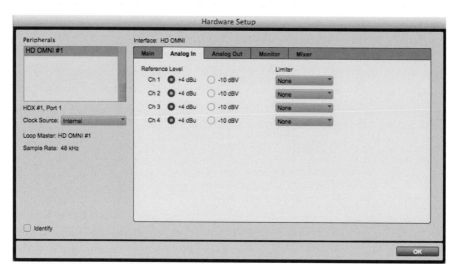

Figure 2.10
HD OMNI Hardware Setup, Analog In page.

Similarly, the Analog Out tab page lets you set reference levels for the outputs – see Figure 2.11.

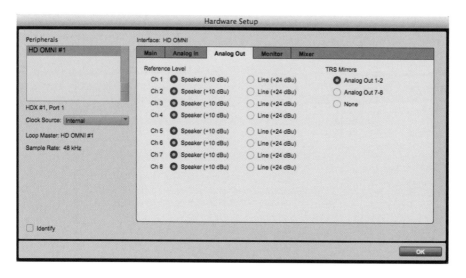

Figure 2.11
HD OMNI Hardware Setup, Analog Out page.

The Mixer page lets you set up the levels for monitoring through the various outputs – see Figure 2.12.

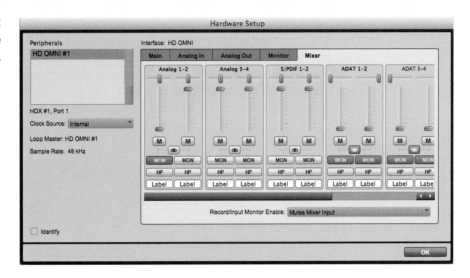

Figure 2.12
HD OMNI Hardware
Setup, Mixer page.

The I/O Setup Window

The I/O Setup window lets you label, format, and assign Pro Tools input, output, insert, and bus audio signal paths for individual sessions and for your specific Pro Tools system. Input, Output, and Insert pages display a graphical representation of the signal routing (the Channel Grid) for physical input and output paths for each connected audio interface with controls that let you route physical inputs and outputs on audio interfaces to Pro Tools input and output channels. Controls for PRE (Mic Preamp) signal paths and Delay Compensation settings for hardware inserts are provided on additional pages within the I/O Setup window.

> **NOTE**
> When you are working in the Mix and Edit windows, signals are routed to and from tracks, sends, and inserts using track Input, Output, Insert, and Send selectors. When you click any of these selectors, the paths created and defined in the I/O Setup are what you will see listed there.

A logical grouping of multiple inputs, outputs, inserts, or busses that has a single name and (channel) format is referred to as a signal 'path' in Pro Tools terminology. These paths can include main paths and sub-paths. An example of a main path would be a master stereo output path with its left and right channels. A sub-path represents a signal path within a main path. For example, a default stereo output bus path consists of two mono sub-paths, left and right. Mono tracks and sends can be routed to either mono sub-path of the stereo output bus path. Multichannel bus paths can have any number of sub-paths.

> **NOTE**
>
> With Avid HDX and HD Native systems, you can define which physical ports on your audio interface are routed to available input and output channels in the I/O Setup. Any changes made here are also reflected in the Hardware Setup, and vice versa.
>
> For Pro Tools systems such as the Mbox Pro and the 003, and for HD MADI, physical outputs are fixed.
>
> For third-party and built-in hardware, click the Launch Setup App button in the Hardware Setup for available configuration options.

The I/O Setup dialog window has six pages that you can access by clicking on the 'tabs' that run across the top of the window, marked Input, Output, Bus, Insert, Mic Preamps, and H/W Insert Delay.

Input

The Input page of the I/O Setup allows you to create and assign Pro Tools Input channels to receive audio from the physical inputs of your audio hardware – see Figure 2.13.

> **NOTE**
>
> Most of the time, you won't need to change anything in the I/O Setup unless you change your system hardware or you want to customize your I/O paths.

If you double-click on the name of an input path, you can rename this with something more meaningful if you prefer, such as AKG C12 instead of Input 1. Also, under the column headed Format, pop-up selectors let you choose mono, stereo (or multichannel with PT HD) formats.

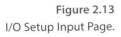

Figure 2.13
I/O Setup Input Page.

For example, I named Pro Tools input and output pair 9-10 'S/P DIF' to make it clear that these were connected to the S/P DIF I/O on the hardware. It is always a good idea to use descriptive names for the Inputs, Outputs, Inserts, and Buses, as this makes things much easier when you are setting up your tracks and mixer configurations during your Pro Tools Session.

> **TIP**
> If you want to return the I/O settings to their initial states, re-setting all the names, formats, and assignments, you can go through the different tabs for Input, Output, Insert, Bus, and so forth, clicking on the Default button for each I/O Setup section to reset these labels to the defaults for your system.

Output

The Output section in the I/O Setup dialog (see Figure 2.14) lets you write your own names for the output signal paths, choose the formats for these, and choose which physical outputs of your audio hardware these output paths are assigned to.

Depending on your system hardware, there are various other setting that you can make. With the HD Omni, for example, you can also choose which physical outputs on your interface are to be used for the Output Meter Path, the Audition Paths, the Default Output Bus, the After Fader Listen (AFL)/Pre Fader Listen (PFL) path, and the AFL/PFL Mute. You can also choose a Default Monitor Format and a 5.1 Path Order for multichannel operation.

Figure 2.14
I/O Setup
Output Page.

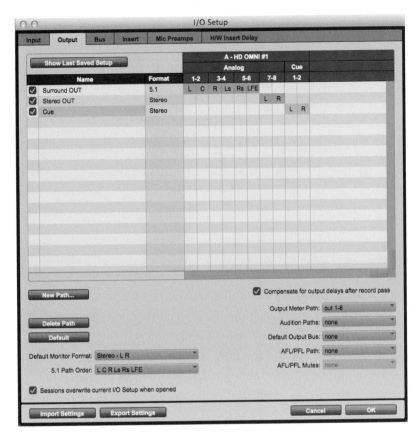

Figure 2.14
I/O Setup
Output Page.

The Output Meter Path selector lets you choose which output or bus paths will be monitored by the Output Meters in the transport.

Pro Tools HD software has a Default Monitor Format pop-up menu that lets you select the default monitor format (Stereo, 5.1, or 7.1) for new Output paths and for when you click the Default button. This setting does not affect existing path definitions or metering – it only specifies channel mapping in new 5.1-format paths.

With HD Native systems, the Output page of the I/O Setup also has an option to enable Low Latency Monitoring, using Outputs 1–2 by default. It also lets you specify any available Output Path for Low Latency Monitoring, and this can be of any channel width from Mono to 7.1.

Bus

The Bus page of the I/O Setup (see Figure 2.15) lets you type in your own bus path names, choose the formats, and map any main bus path to any of the available output paths of the same channel width or greater. So, for example, a mono bus can be mapped to a mono output path, a stereo bus can be mapped to a stereo output path, and a 5.1 surround bus can be mapped to a 5.1 surround output path. The bus page also lets you map output busses to output paths.

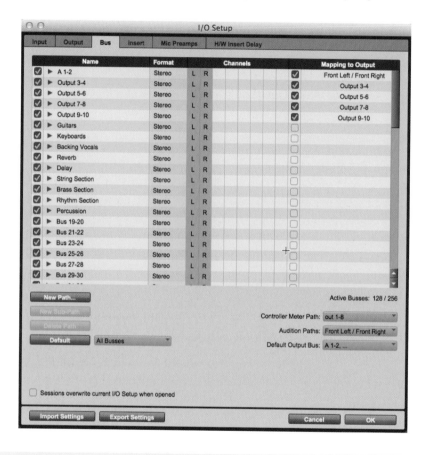

Figure 2.15

I/O Setup Bus Page.

TIP

You can configure these output paths – naming, formatting, and assigning them to physical outputs – in the Output page of the I/O Setup.

NOTE

The Controller Meter Path selector, which is only available in the Bus page of the I/O Setup, determines the path displayed across the Output meters of D-Control or D-Command work surfaces.

Pro Tools provides up to 256 internal mix busses that can be used to route audio from track outputs and sends to other track inputs or to plug-in side-chains. Busses are typically used with the Sends in the Pro Tools Mix window to route audio from one or more audio tracks to an Auxiliary Input track used as a Reverb or Effects return (to the mix) channel. Another popular use is to create sub-mixes of groups of tracks containing Backing Vocals, Strings, Brass, Guitars, Keyboards – or whatever you find useful – by routing the outputs of these tracks via internal busses to Auxiliary Input tracks. Busses can also be used to route audio via Auxiliary tracks to physical outputs on an audio interface to use as Cue or Headphone mixes.

Insert

You can create and edit hardware insert signal paths for the Pro Tools mixer using the Insert page of the I/O Setup, (see Figure 2.16), naming these appropriately and choosing which pairs of inputs and outputs to use for these on your audio interface. So, for example, if you want to be able to use some external 'outboard' effects devices, such as a Lexicon 224 reverb or even a vintage EMT 140 'echo' plate, you would connect the inputs and outputs of these devices to available inputs and outputs on your Pro Tools audio interface.

Figure 2.16
Part of I/O Setup
Insert Page.

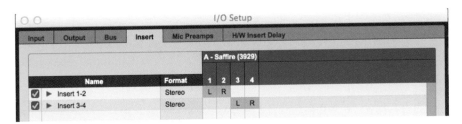

Mic Preamps

If you are using an Avid PRE, you can map signal paths for one or more of the PRE's microphone preamplifiers to your audio interface(s) using the Mic Preamps page of the I/O Setup.

H/W Insert Delay

If you are using any external hardware devices, such as a reverb unit or a compressor, you can set a specific amount of Hardware Insert Delay Compensation, in milliseconds, for each external device to compensate for the delay (latency) that will occur between sending audio from Pro Tools to the device and Pro Tools receiving audio back from the device – see Figure 2.17.

When the hardware insert is in use and Delay Compensation is enabled, these delay times will be used by the Delay Compensation Engine to time-align the input paths.

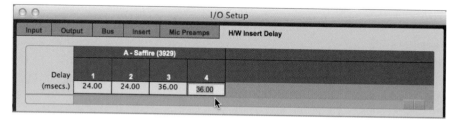

Figure 2.17
Part of I/O Setup H/W Insert Delay Page.

I/O Setup Options

Pro Tools systems have several additional I/O Setup options depending on which page of the I/O Setup you are viewing. These include default signal routing for metering and auditioning and default track layout for multichannel mix formats.

I/O Delay Compensation

Pro Tools|HDX and Pro Tools|HD Native Systems provide two options for compensating for input and output latency (due to any inherent latency in the analog-to-digital and digital-to-analog converters of the audio interface) after recording.

A 'Compensation for Input Delays After Record Pass' option is available in the I/O Setup Input page – see Figure 2.18. This provides automatic compensation for any analog or digital input delay with Avid HD interfaces, so you should enable this option whenever you are recording.

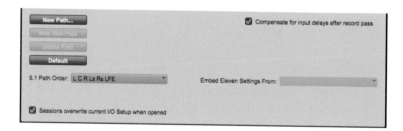

Figure 2.18
Compensation for Input Delays After Record Pass option in the I/O Setup Input page.

Similarly, in the I/O Setup Output page, there is a 'Compensation for Output Delays After Record Pass' option that provides automatic compensation for any analog or digital output delay with Avid HD audio interfaces – see Figure 2.19. You should enable this option whenever you are synchronized to an external clock source.

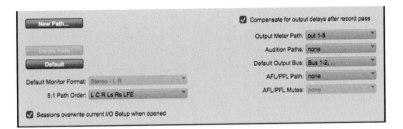

Figure 2.19
Compensation for
Output Delays After
Record Pass option
in the I/O Setup
Output page.

NOTE

When recording from a digital source, both the Compensation for Input Delays After Record Pass and the Compensation for Output Delays After Record Pass options must be enabled.

Audition Paths

You can specify the output path to be used as the 'Audition Path' to playback files and clips when you audition or 'preview' these from the Clip List or Workspace browsers, and when previewing AudioSuite processes. Pro Tools assigns as default Audition Path, the first available main Output path of the corresponding format. If you prefer, you can select a specific Audition Path using the Audition Paths pop-up selector, which is available in the Output and Bus pages of the I/O Setup.

Default Output Bus

You can also specify the default output bus path assignment for new tracks, in each available format using the Default Output Bus selector, which is available in the Output and Bus pages of the I/O Setup. The Default Output Bus can also be set for internal mix bus paths.

AFL/PFL Path

When you are using Pro Tools HD with Avid HDX or HD Native Hardware, there is an AFL/PFL Path selector available in the Output and Bus pages of the I/O Setup. Tracks soloed in AFL or PFL Solo mode are played back via the current AFL/PFL Path that you have set using the AFL/PFL Path selector. Be aware that if you select 'None' as the AFL/PFL path, this disables the AFL and PFL solo modes.

NOTE

In the Mix or Edit window, you can set separate master playback levels for all AFL and PFL solos. Here's how this works: choose the AFL or PFL solo mode from the Options menu Solo Mode sub-menu first. Next, Command-click (Mac) or Control-click (Windows) a Solo button on any track in the Mix or Edit window. Adjust the AFL/PFL Path fader then click on the new fader position (or press Esc) to close the fader display. To set the AFL/PFL Path level to 0 dB, Command-Control-click (Mac) or Control-Start-click (Windows) on any Solo button.

AFL/PFL Mutes (Output Path)

The AFL/PFL Mutes selector becomes available in the Output and Bus pages of the I/O Setup when you are not using a D-Control or D-Command work surface. This allows you to mute the normal Pro Tools output path when you send a signal to the AFL/PFL Path. To set which output path will be muted when tracks are soloed in AFL or PFL Solo mode, select a path from the AFL/PFL Mutes (Output Path) pop-up selector.

5.1 Path Order

If you are using Pro Tools HD, a 5.1 Path Order selector is available in the Input, Output, and Insert pages of the I/O Setup. This lets you specify the default track layout for all new 5.1 format paths you create, and you can choose between C|24/Film, SMPTE/ITU, DTS/ProControl Monitoring, or D-Command/D-Control.

I/O Setup Rules

Path configurations, Input, Output, Insert, and bus names, and channel widths in the I/O Setup are saved as I/O settings with both the session and the system and can be recalled from either.

When you take a session from one system to another, track and send assignments are maintained within the session and, where possible, Pro Tools automatically remaps the session's output busses to the output paths of the system on which the session is being opened.

A checkbox near the bottom of the Input, Output, Bus, and Insert pages of the I/O Setup window lets you choose whether the I/O settings saved with the session will overwrite the I/O settings saved with the system when you open the session. When this option is enabled, which is the default, Pro Tools recalls these settings from the session rather than the system. This option is the best choice when exchanging sessions with systems running Pro Tools 8.0.x and lower. When this option is disabled, Pro Tools recalls these settings from the system, which is the best choice when exchanging sessions among different systems running Pro Tools 8.1 or higher.

When you create a new session, you can specify which I/O Settings to use in the I/O Settings pop-up menu in the Quick Start and New Session dialogs. For example, you can use the factory installed default settings, the 'Last Used' setting, or one of any available custom I/O settings files.

If any changes are made to the I/O Setup, these changes are automatically saved to the I/O Settings folder as the Last Used settings file when the I/O Setup is closed.

Custom I/O Settings files can be created by changing I/O Setup settings and then exporting the settings. These I/O settings can then be restored by importing them into a system.

I/O settings can be imported and exported for use with sessions shared between systems, and these can be imported either before or after you open a session. I/O Settings are only imported for the current page of the I/O Setup – helping you to avoid overwriting any I/O settings you have made on the other pages. When you export I/O settings, on the other hand, all the pages of the I/O Setup are exported, so that all the latest changes you have made are preserved.

Session Interchange Rules

Every time you save a Pro Tools session, the software finds an identifying number, or system ID, for the computer system you are using (based on the computer's MAC address) and saves this within the session file. If Pro Tools finds a matching system ID the next time you open this session, this means you are using the same computer that the session was created on, and the output paths are immediately restored with no need for any reconfiguration.

Whenever a Pro Tools session is opened on a system, Pro Tools attempts to automatically remap output busses. If Pro Tools does not find a matching system ID, it tries to find matching, identical, Path Names and Formats to remap output busses to, and if it cannot find these, it will look for paths of the same format to remap to. If it cannot find any matches for any output buses, these will be opened as 'Inactive' and you will have to manually remap these paths to active output paths.

Getting Started

The Quickstart Dialog

When you launch Pro Tools, the QuickStart dialog appears (unless you have previously deselected this option in the Warnings & Dialogs section of the Display Preferences or in the Quickstart Dialog itself). Using this dialog, you can create a blank session, create a new session from a template, open any of the last 10 previously opened sessions, or open any existing session.

> **TIP**
>
> If you click on the Session Parameters arrow in the QuickStart dialog, you can choose a different audio file type, bit depth, or sample rate for your new session.

A selection of templates is provided in the QuickStart dialog (see Figure 2.20) containing useful Session configurations – and you can easily create your own if you prefer. Simply create a new Pro Tools session, configure it however you like, and choose 'Save As Template' from the File menu. For songwriting, you might just have one mono track for guitar, a second mono track for voice, a stereo Instrument plug-in for drums or percussion, an Auxiliary Input with a reverb plug-in inserted, and a stereo Master Fader.

In the Save Session Template dialog, you can choose 'Install Template In System', which installs the template file in the system folder referenced by the Pro Tools Session Quick Start dialog.

Alternatively, you can choose 'Select Location For Template', which lets you select any other location, in which case the session template won't appear in the Pro Tools Session Quick Start dialog – but you can simply open this file to start a new session from this template.

Figure 2.20
The Quick Start
dialog.

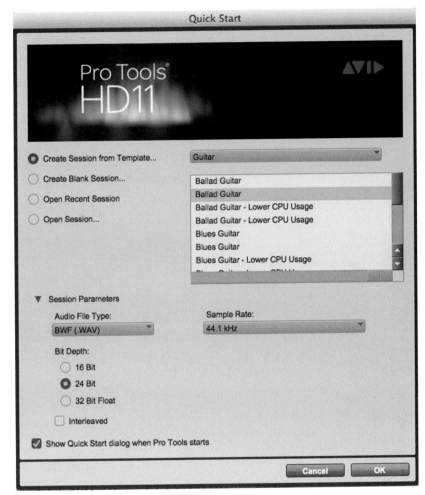

TIP

You might prefer this second method if you want to keep the template with the project that you are working on and intend to move this to another system at some point during the project.

When the Include Media option is enabled, any audio, MIDI, or video media in the session is included in the template. This is useful if you want to use some standard media elements, such as a particular sound effect or drum loop, in a series of related Pro Tools Sessions.

Opening Sessions with Plug-Ins Deactivated

If you are using a lot of plug-ins in a particular session, these can take a long time to load. If you simply want to open the session to check if this is the one you want to work with, or to make some changes that you will not need the plug-ins for, Pro Tools provides a convenient way to open sessions with all of the session's plug-ins set to inactive.

From the File Menu, choose Open. When the Open Session dialog appears, first locate and select the session you want, then Shift-click Open. The session will now open with the plug-ins deactivated.

If you decide that you do want to work with this file with the plug-ins loaded and activated, you can either choose Revert To Saved from the File menu or choose Open Recent from the File menu and select the most recent session in the sub-menu.

> **TIP**
> There is a convenient keyboard command that lets you quickly re-open the most recently opened session: Command-Shift-O (Mac) or Control-Shift-O (Windows). This is very useful if you have just closed this file and want to immediately open it again, perhaps because you forgot to do something.

Pro Tools Sessions

So what is a 'session'? Well, this is the name given to the project document that Pro Tools creates when you create and configure a new session. This document, which has the file extension '.ptx' when it is saved to disk, contains all your edit and mix information, and it references any audio, MIDI, video, or other files associated with this project. The session file does not actually contain any audio or video media files – these files are quite separate – and only one session file can be open at any one time.

> **TIP**
> You can always make changes to a session and save those changes in a new session file to create multiple versions of a session or to back up your editing and mixing work.

When you have created or opened a Pro Tools session, you can work on your recording, editing, mixing or post-production session, importing, exporting, creating or referencing audio, video, or MIDI 'media' files. MIDI data can be created directly within the session and stored in the session document, or can be imported or exported as standard MIDI files, while audio and video files recorded or imported into a Pro Tools session are always stored separately from the Pro Tools session file in dedicated Audio Files and Video Files folders within the Pro Tools Session folder.

> **NOTE**
> Audio files are listed in the Pro Tools Clip List and can appear in an audio track. A section of an audio file can be defined as a clip. Video files can be created in (or copied to) the Video Files folder in the session folder. However, in most cases, Pro Tools references video files that have been captured by another application, such as Avid Media Composer.

Tracks

Within a Pro Tools session, video, audio, MIDI, and automation data can be recorded and edited on video, audio, MIDI, or Instrument 'tracks' that can be viewed in the Mix and Edit windows.

Auxiliary Input tracks can be used route additional external inputs (or internal sources entering via internal busses) out to the various outputs, or out via internal busses to other destinations. These Auxiliary Inputs are often used to create sub-mixes or used as Effects Returns with reverb, delays, or other effects inserted.

Master Faders are normally used to control the level and panning of audio being routed to physical outputs, although these have other uses that will be discussed later in this book.

Pro Tools HD also has VCA Master tracks that allow you to control other tracks in a Mix Group that has been assigned to the VCA Master.

> **TIP**
> Pro Tools 11 offers a great new way to create new tracks by double-clicking in blank space below tracks in the Edit window, or below or to the right of tracks in the Mix window, or in the empty area below any current tracks in

(Continued)

the Tracks list. Double-clicking in any of these blank areas will add a new track of the same type (Audio, Auxiliary Input, Instrument, VCA Master, MIDI, or Master Fader) and channel width as the last new track that you added. If there are no tracks in the session yet, a stereo audio track will be created by default.

If you want to ensure that the track type added will be an audio track of the same channel width as the last new track, then hold the Command key (Mac) or Control key (Windows) as you double-click.

If you want to specifically create a new Auxiliary track of the same channel width as the last new track, hold the Control key (Mac) or Start key (Windows) as you double-click. If no tracks exist in the session, a stereo Auxiliary Input track is created by default.

To add a new stereo Instrument track, hold the Option (Mac) or Alt (Windows) key.

To specifically add a new Master Fader track of the same channel width as the last new track, hold the Shift key as you double-click, and, again, if no tracks exist in the session, a stereo Master Fader track will be created by default.

Clips

A segment of audio, MIDI, or video data is referred to as a 'clip'. So, a clip could contain a drum loop, a guitar riff, a verse of a song, a recording take, a sound effect, and some dialog – or an entire sound file. A clip can also have associated automation data. Video tracks (on Mac or Windows) let you work with QuickTime movies or VC-1 video files (on Windows), or with Avid video – although you cannot mix these formats on a single track.

In a typical Pro Tools session, clips are created from audio files or MIDI data, and arranged in audio and MIDI track 'playlists' along the timeline in the Edit window. Clips can also be grouped (to form a 'clip of clips') and they can be made to act like 'loops' by being repeated end-to-end along the timeline in the Edit window for as long as you would like the 'loop' to last. Using these techniques, you can re-arrange sections or entire songs, and assemble tracks using material from multiple takes.

Playlists

A 'playlist' is simply a sequence of clips arranged on an audio, MIDI, or video track, and tracks can have both Edit playlists and Automation playlists.

For example, on an audio track, you might have recorded a guitar part that lasts for four bars of an eight bar intro, followed by a second guitar part that plays in the bridge sections, with a third guitar part that plays during the end choruses. These parts may have been played by three different guitarists and recorded at different times during your recording project, and the three parts will each be contained within separate Clips. When you come to mixing the track, you may decide to have copies of the guitar part from the end choruses play during the earlier choruses and maybe you will decide to add a chorus effect to the guitar parts that play in the bridge sections. All this can be done using a single Edit playlist with an Automation playlist to control the effects. Because you can use a copy of the audio clip to play one guitar part in different locations, this doesn't use any additional disk space, as the copy of the clip simply plays the original guitar in the new locations. It is also very easy to apply different effects to an original piece of audio, and then play this back in different places.

Perhaps the most powerful feature of playlists is that you can create any number of alternate playlists that you can choose from within each track. An example of how you might use these within a single audio track is when you are recording, say, a saxophone solo. You could record the musician playing several different versions of the solo into different playlists within the same track, then choose between these later – maybe using part of one 'take' followed by part of a different 'take', until you have created the perfect combination of notes.

NOTE
Each audio, Auxiliary Input, Instrument, Master Fader, and VCA track also has a single set of automation playlists. Automation playlists can include volume, pan, mute, and each automation-enabled control for the insert and send assignments on that track. MIDI controller data on Instrument and MIDI tracks is always included as part of the track playlist.

Channels vs. Tracks

In Pro Tools terminology, 'tracks' are used to record audio (or MIDI) to your hard drive. These tracks appear in the Edit window and in the Mix window 'channel' strips.

Rather confusingly, what you might expect would be referred to as 'channels' in the Mix window are referred to as Audio Tracks, Video Tracks, MIDI Tracks, Instrument Tracks, Auxiliary Inputs, VCA Master, and Master Faders in Pro Tools.

Also, Pro Tools uses the term 'channel' to refer to an actual physical input or output connection on whichever interface or interfaces you are using, so the documentation talks about an interface having 16 channels of audio input and output, for example.

All this is straightforward enough to understand – but it is a little unusual when you are used to a tape recorder having tracks and a mixer having channels. It can be a little 'jarring' mentally to talk about adjusting the mixer's 'track' controls rather than the mixer's 'channel' controls until you get used to the idea. But the documentation still talks about 'channel' strips in the Pro Tools Mix window – each of which corresponds to a track in a Pro Tools session.

Voices and 'Voiceable' Tracks

The concept of 'voices' in Pro Tools needs to be thoroughly understood. The number of playback and recording 'voices' is the number of unique simultaneous playback and record tracks on your system – a bit like the concept of polyphony in a synthesizer, which dictates the maximum number of simultaneously playable notes.

Every Pro Tools system has a maximum number of voices, depending on the hardware it uses, that defines how many audio tracks you can play back or record at the same time. Pro Tools terminology, somewhat confusingly, also refers to 'voiceable' or 'virtual' tracks, when describing the maximum number of audio tracks that the software can support in a session.

For example, a Pro Tools HDX system with one HDX card can support up to 768 'voiceable' or 'virtual' tracks (using Avid's terminology for these) – but you cannot record to these all at once. And you can't play all these back at once either. You can only record and play back up to the limit of the available 'voices' within your system, and this depends on the type of card or cards you have in your system and the sample rate you are working at. An HDX system with one card, for example, provides a maximum of 256 voices at 48 kHz and the older Pro Tools HD cards supported a maximum of 192 voices at 48 kHz – yet Pro Tools HD software running on either of these cards supports up to 768 voiceable tracks! To put this another way: the total number of 'voiceable' tracks is the maximum number of audio tracks that can share the available voices on your system.

Are you still feeling confused? Well, here's the thing: the Pro Tools software lets you create many more audio tracks than there are available voices, and save these in your session, but you just cannot record to or play back more tracks than there are available voices at any one time. With an HDX card, you could record up to 256 tracks at once, then record a further 256 tracks, and then a further 256 tracks for a total of 768 tracks. When you playback this session using just one HDX card, you would only hear the 256 tracks with the highest track playback priority – the others would not be heard. But if you have three HDX cards, you could hear all 768 tracks!

> **NOTE**
> By way of comparison, standard Pro Tools systems let you play back up to 96 simultaneous stereo or mono tracks, but can only record up to 32 tracks simultaneously.

Here are some relevant facts and figures to further clarify the situation: Pro Tools HD can open sessions with up to 768 audio tracks, but any audio tracks beyond that system's voiceable track limit will be automatically set to Voice Off. Each HDX card lets you record and playback up to 256 tracks at once at 44.1/48 kHz, you can use up to 3 cards in a system, and each HDX card supports a maximum of 64 channels of I/O. If your requirements are somewhat humbler, an HD Native system may suffice. HD Native hardware lets you record and play back up to 256 tracks at once at 44.1/48 kHz and supports a maximum of 64 channels of I/O, but, unlike the HDX cards, doesn't have on-board DSP processing. How many tracks you can actually record at once will also depend on how many inputs you have available via your audio interface or interfaces. For example, Avid interfaces offer a range of configurations: HD OMNI has 8 inputs and 8 outputs, HD I/O has 16 inputs and 16 outputs, and HD MADI has 64 inputs and 64 outputs.

A big advantage that Pro Tools HDX cards have over the previous-generation Pro Tools HD cards is that they use dedicated Field Programmable Gate Arrray (FPGA) chips to provide the 256 voices on each card in addition to the DSP chips used to support plug-in processing and mixing. As a consequence, the number of tracks that you use in a session does not reduce the amount of DSP available for plug-in processing and mixing and the maximum voice count is always available.

On Pro Tools HDX systems, the number of available voices will depend on how many of these cards you are using in your system – the more cards you have, the more voices you can use. On all other Pro Tools systems, the maximum number of voices may be limited by the host-processing power of your computer – so you will need to use a powerful computer for optimum results.

Using Up Voices

Each audio channel for each track in your Pro Tools session uses a single voice. So, for a mono audio track, a single voice is used; stereo and multi-channel tracks take up one voice per channel. However, you do need to bear in mind that if you are using Punch Recording, two voices are needed for every single audio channel (one for playback and one for recording on punch in and out).

Also, with Pro Tools|HDX systems, when you insert a Native (host-based) plug-in, this may cause additional latency and will take up two additional voices per channel (one voice for input and one voice for output) when inserted on an Auxiliary Input or Master Fader track; or when inserted on an Instrument track that does not contain an instrument plug-in; or when inserted after a DSP plug-in on any kind of track.

For example, the initial insert of a host-based plugin on a mono Auxiliary Input track uses two voices (one channel with two voices), while the initial insert of that plug-in on a stereo Auxiliary Input track uses four voices (two channels with two voices each). Subsequent host-based plug-ins on the same track do not take up additional voices unless a DSP plugin is inserted between other host-based plug-ins.

> **NOTE**
> You should always avoid inserting DSP plug-ins between host-based Native plug-ins on any kind of track as this will cause unnecessary voice usage and may cause additional latency.

There are various other scenarios in which additional voices are used: when you select multiple track outputs for a track, one voice is used for each output; and when you select an AFL/PFL Path output in the Output tab of the I/O Setup dialog, one voice is used for each channel. Also, when you use the external key side-chain of a host-based plug-in on a track, one additional voice is used.

Figure 2.29
Window Layout
configuration.

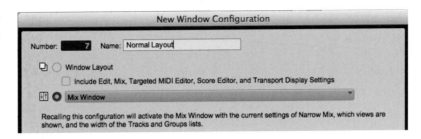

The Transport Window Display Settings option stores all window display settings for the Transport window, including the Counters, MIDI controls, and Expanded view settings, and lets you recall the Transport window with these settings when it is closed.

The Score Editor Window Display Settings option stores all the current window display settings for the Score Editor window.

The Targeted MIDI Editor Window Display Settings option stores the current automation panes and heights, and the current settings of which rulers and rows are being shown in the targeted MIDI Editor window.

TIP

If you just want to be able to recall the size, location, and order of a particular set of windows that you like to work with, but without recalling the display settings within those windows, you can simply store the Window Layout with the 'Include ... Display Settings' option de-selected in the New Window Configuration dialog.

The Universe View

Pro Tools has a navigation feature called the Universe View that runs along the top of the Edit window. This displays an overview of the entire session, showing a compact representation of all the audio and MIDI data on tracks that are not hidden.

Audio, MIDI, and video clips on tracks are represented by horizontal lines that are the same colours as the clips on the tracks and each channel in a stereo or multichannel audio track is represented individually. Since Auxiliary Input, Master Fader, and VCA Master tracks do not contain audio or MIDI clips, they are displayed as blank areas in the Universe view – see Figure 2.30.

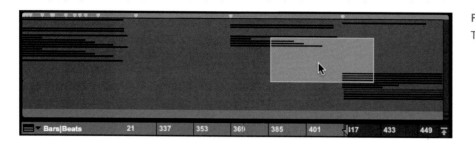

Figure 2.30
The Universe View.

To make material visible within the confines of the Edit window, you often need to use the vertical and horizontal scroll bars so that you can see what you need to see. The Universe View provides an alternative way of positioning material in the Edit window so that it is visible.

A white rectangular area within the Universe View represents what is visible in the Tracks area of the Edit window. You can click on and drag this white rectangle to move the area that it encompasses (or 'frames') and the material displayed in the Edit window will follow. This can be a much faster method than using the scroll bars to achieve the same result.

The framed area updates to reflect what is happening in the Edit window, so when all the tracks are made visible in the Edit window and the session is zoomed all the way out with all clips visible in the Edit window, the framed area encloses the whole of the Universe window. During playback, if the Edit window is set to scroll, the framed area in the Universe view will scroll to keep in step with this.

Pro Tools also has a Video Universe window accessible from the Window menu that lets you view, navigate, zoom, and select video clips on the main video track.

The Main Windows

The Edit and Mix windows are the main Pro Tools work areas. Depending on which phase of your project you are in or what type of project you are working on, you may prefer to work with just the Mix or just the Edit window.

You can use the View Menu to choose what will be presented to you in the Edit and Mix windows. So, for example, if you reveal the Instruments, I/O controls, Inserts, and Sends in the Edit window, you can mostly work in just the Edit window, which some people find easier than working in two windows (see Figure 2.31).

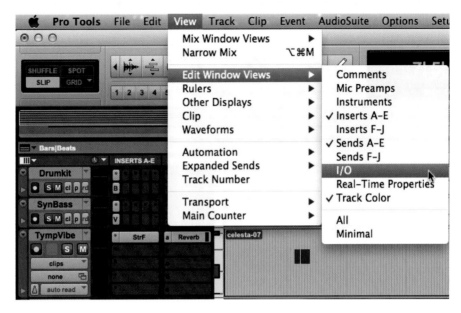

Figure 2.31
Revealing the I/O
controls in the Edit
window.

Organising the Transport Window

The Transport Window takes its name from the name given to the controls that were used on tape recorders to control the tape transport – that is, the mechanism that caused the tape to be transported along its path across the tape heads. Presumably, Avid chose this name to make recording engineers and producers who started out using tape recorders feel more comfortable with their digital audio recorder. Interestingly, a hard disk drive is like a cross between a tape recorder and a vinyl disc recorder: like a vinyl disc recorder, it has a rotating platter and a read/write head on a movable arm (or several of these), and like a tape recorder, it records the audio as patterns of magnetization on a magnetic medium – an analogue process. But as far as the computer is concerned, the audio is recorded and played back digitally as a series of 1's and 0's from a file on disk.

Clearly, the Transport Window is an important part of any Pro Tools system – allowing you to control recording and playback of audio (and MIDI) – so you need to make sure that you are totally familiar with all its controls and functions.

In standard view, the Transport window just has a row of buttons providing controls for Online, Return To Zero, Rewind, Stop, Play, Fast Forward, Go to End, and Record. The Transport sub-menu that you can access from within the View menu lets you display Synchronization controls, MIDI Controls, and Counters in the Transport window as well. Synchronization controls include two buttons: one to enable Online timecode synchronization and one to Generate MIDI Time Code (MTC) using a SYNC peripheral.

There are four MIDI control buttons: Wait for Note, Metronome, MIDI Merge, and Conductor. Wait for Note does what it says: Pro Tools won't play or record until you send it a MIDI note. This is useful when you are controlling Pro Tools yourself and are playing a MIDI instrument such as a keyboard. The Metronome button enables the click, assuming you have set up a click track correctly. MIDI Merge lets you record new MIDI data onto a track that already contains MIDI data, merging all the data together. The Conductor button lets you enable or disable the Conductor track that contains tempo information for your Pro Tools session.

The Main Counter shows your choice of Bars:Beats, Mins:Secs, Samples, Timecode, Timecode 2, or Feet + Frames and you can choose between these using the pop-up selector that appears when you click and hold the small downwards-pointing arrow at the right of the counter display.

You can also switch the Transport window to its Expanded display – see Figure 2.32. This adds a Sub-counter underneath the main counter, along with Count Off, Meter, and Tempo controls above the MIDI controls. You can either enter the tempo manually here or you can enable the Tempo track by clicking on the Conductor button.

Underneath the transport controls, the Expanded Transport window adds Pre- and Post-roll settings, along with Start, End, and Length indicators for making Timeline selections.

Figure 2.32
Transport Window expanded to show the Counters and MIDI Controls with the pop-up selector for these revealed at the far right.

TIP
To start and stop playback, simply press the Spacebar on your computer keyboard.

Transport Preferences

To be fully in control of Pro Tools, you need to understand the effects that the various Transport Preferences (see Figure 2.33) settings have.

If 'Timeline Insertion/Play Start Marker Follows Playback' is selected, when you stop playback the Timeline Insertion and the Play Start Marker, both move to the point on the Timeline where playback stops. Otherwise, when this option is deselected, the Timeline Insertion and Play Start Marker will return to the point on the Timeline from which playback began.

Figure 2.33
Pro Tools HD
Transport Preferences.

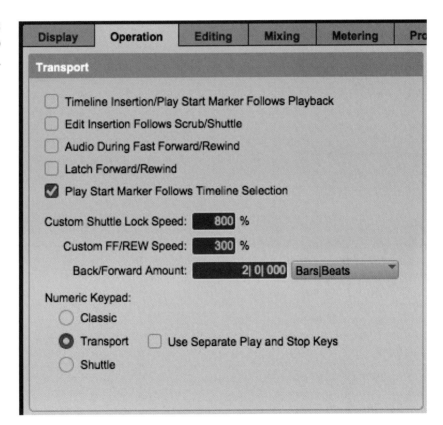

NOTE

This is such an important setting that there is a dedicated button on the Toolbar, marked 'Insertion Follows Playback', that lets you toggle this option on and off.

When 'Edit Insertion Follows Scrub/Shuttle' is selected, the edit cursor automatically locates to the point on the Timeline where you stop scrubbing the audio.

When 'Audio During Fast Forward/Rewind' is selected, you will hear the audio playing back during fast forward or rewind.

When 'Latch Forward/Rewind' is selected, Fast Forward and Rewind will latch and continue to operate until you start or stop playback. When this preference is disabled, the Fast Forward and Rewind will only last as long as you hold down the mouse after clicking either button on the Transport (or hold down the corresponding switch on a Control Surface).

When 'Play Start Marker Follows Timeline Selection' is enabled, the Play Start Marker snaps to the Timeline Selection Start Marker when you move the Timeline Selection, or when you draw a new Timeline Selection, or when you adjust the Timeline Selection Start. When this preference is disabled, the Play Start Marker doesn't move with the Timeline selection.

Pro Tools HDX and HD Native systems have an extra option in the Transport Preferences: 'Reserve Voices for Preview in Context'.

When 'Reserve Voices For Preview In Context' is enabled, Pro Tools reserves the appropriate number of voices for preview in context (the feature that allows previewing audio files in Workspace browsers during session playback).

> **NOTE**
> When you are reserving voices for Preview in Context, these are no longer available to use for other purposes, because the number of available voices is inevitably reduced by the channel width of the selected Audition Paths on the Output page of the I/O Setup window. For example, if the number of playback voices is set to 48 in the Playback Engine, and you have a stereo audition path selected in the I/O Setup, only 46 voices will be available for tracks. If you have a 5.1 audition path, only 42 voices will be available. When this option is disabled, you will not be able to preview in context if there are not enough available voices.

The Custom Shuttle Lock Speed preference (Pro Tools HD only) allows you to customize the highest fast-forward Shuttle Lock speed (key 9) in Transport or Classic numeric keypad modes. The range for this setting is 50% to 800%.

The Custom FF/REW Speed preference lets you set the speed for fast forward and rewind in relation to normal playback and ranges from 100% to 3,200%.

You can use the Back/Forward Amount (Pro Tools HD only) to set the default length of Back, Back and Play, Forward, and Forward and Play. The timebase of the Back/Forward Amount settings follows the Main Time Scale by default or you can deselect Follow Main Time Scale and select another timebase format: Bars:Beats, Min:Sec, Time Code, Feet+Frames, or Samples.

When 'Use Separate Play and Stop Keys' is enabled, you can start playback with the Enter key and stop playback with the Zero (0) key on the numeric keypad. 'Use Separate Play and Stop Keys' is only available when you are using the Numeric Keypad's Transport mode.

> **TIP**
> Setting up Pro Tools to use the Enter and '0' keys on the numeric keypad to start and stop playback makes Pro Tools behave like Cubase and other popular MIDI + Audio software applications and is particularly useful when you want to quickly audition loop transitions.

> **NOTE**
> When 'Use Separate Play and Stop Keys' is enabled, this option overrides using the Enter key to add Memory Location markers, so if you do need to add a Memory Location marker while this is active, you need to press Period (.) before you press Enter on the numeric keypad.

Scrub/Shuttle and the Numeric Keypad Modes

To understand which Numeric Keypad mode to choose, you need to understand the Scrub/Shuttle features first.

Using the Scrubber Tool

Scrubbing is a technique borrowed from tape editing, where the tape was moved back and forth past the playback head by hand while listening to the audio to find a suitable edit point for cutting or splicing. The Scrubber tool in Pro Tools emulates this procedure, allowing you to 'scrub' up to two tracks of audio in the Edit Window (MIDI and Instrument tracks cannot be scrubbed).

To scrub a single audio track, select the Scrubber tool then point, click and drag within the track, moving left for reverse or right for forward – see Figure 2.34.

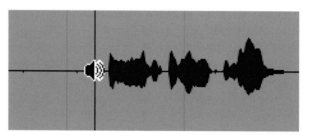

Figure 2.34
Scrubbing by dragging within a single track.

To scrub two audio tracks, drag between the adjacent tracks or scrub within a selection that contains multiple tracks – see Figure 2.35.

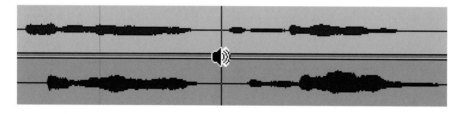

Figure 2.35
Scrubbing by dragging between two adjacent tracks.

If the 'Edit Insertion Follows Scrub/Shuttle' option is enabled in the Operation Preferences page, when you stop scrubbing the edit cursor will automatically locate to the point where you stopped. It makes most sense to enable this option when you are using the Scrubber tool because you will usually want to insert the edit cursor at this point to make your edit.

To make the scrub feature even more like tape editing, Pro Tools HD allows you to set the Scrolling option to Continuous or Center Playhead. With either of these scrolling modes, wherever you click in a track's playlist with the Scrubber tool activated, Pro Tools scrolls so that the Playhead is centred in the Edit window. Then, as you scrub the audio, it moves past the Playhead, which remains stationary and centred – more like moving tape back and forth past a fixed play head – see Figure 2.36.

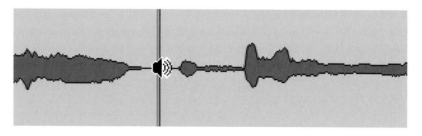

Figure 2.36
Scrubbing with Center Playhead scrolling enabled.

To get used to using the Scrubber tool is easy enough. Just move the mouse back and forth over the audio where you are trying to find the correct edit point, moving progressively more slowly until you have identified the spot. This can be much easier to control if you are using a scrub wheel on a control surface.

Keep in mind that the speed of playback of the scrubbed audio depends on how fast and how far you drag. Also, the resolution for the Scrubber is dependent on the zoom factor for the scrubbed track, so you may need to zoom in or out for best results.

> **TIP**
> You can temporarily switch the Selector tool to the Scrubber tool by Control-clicking (Mac) or Start-clicking (Windows). For finer resolution, Command-Control-click (Mac) or Control-Start-click (Windows).

Scrub/Shuttle Mode

The Scrubber tool only lets you scrub at normal playback speed or slower. In Scrub/Shuttle mode, the Scrubber tool also lets you scrub at faster speeds. This is useful when you are searching through larger ranges of audio.

To scrub in Shuttle mode, simply hold down the Option key (Mac) or Alt key (Windows), while you drag within the track with the Scrubber tool selected – see Figure 2.37. The Fast Forward and Rewind buttons in the Transport window will engage to indicate that you are in fast forward/rewind Shuttle mode – see Figure 2.38.

Figure 2.37
Scrub/Shuttle mode showing the Fast Forward and Rewind buttons in the Transport window engaged.

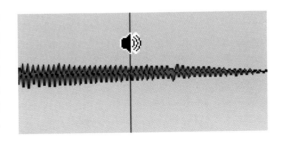

Figure 2.38
Fast Forward and Rewind buttons both engaged to indicate fast forward/rewind Scrub Shuttle mode.

Shuttle Lock Mode

Shuttle Lock mode lets you shuttle forwards or backwards through your audio while controlling the speed and direction using the computer keyboard instead of using the mouse or scrub wheel.

To enter Shuttle Lock mode, hold down the Control key (Mac) or Start key (Windows) and press a number on the numeric keypad. You can let go of the modifier key once playback has started and you can reverse the direction of playback by pressing the Minus (–) or Plus (+) keys on the numeric keypad.

To change the speed of playback, hold down the modifier key again and press a different number on the numeric keypad: 5 is normal speed, numbers from 6 up to 9 provide increasingly faster fast-forward speeds, and numbers from 4 down to 1 provide progressively faster rewind speeds (4 is the slowest rewind Shuttle Lock speed, and 1 is the fastest).

To stop playback, press Control-0 (Mac) or Start-0 (Windows) and to exit Shuttle Lock mode, just press Escape or the Spacebar.

> **NOTE**
> As with all the Scrub/Shuttle modes, if multiple tracks are selected, only the first two tracks will be heard.

> **TIP**
> You can also operate the Transport using the numeric keypad. Press 0 to Play/Stop, 1 to Rewind, 2 to Fast Forward, 3 to enable Record, 4 to turn Loop Playback mode on/off, 5 to turn Loop Record mode on/off, 6 to turn QuickPunch mode on/off, 7 to turn the Click on/off, 8 to turn the Countoff on/off, and 9 to enable MIDI Merge/Replace mode.

You can recall Memory Locations (markers) by typing Period (.), the Memory Location number, and Period (.) again – *using your computer's numeric keypad.*

The Numeric Keypad Shuttle Mode

The numeric keypad Shuttle mode provides an alternative to using Shuttle Lock mode when you want to control Scrub/Shuttle using the computer keyboard instead of the mouse/scrub wheel.

Shuttle mode works quite differently from the two Shuttle Lock modes. The main difference is that to play the current Edit selection, you must press and hold the keys on the numeric keypad: playback stops as soon as you let go of the keys. And you don't have to be in Scrub mode to use the Shuttle feature, because Scrub mode is automatically engaged as soon as you press the relevant keys on the numeric keypad. Also, in this mode, pre- and post-roll are ignored.

Various playback speeds are available in both forward and reverse. Press key 1 to Rewind at quarter-speed, key 3 to go Forward at quarter-speed, key 4 to Rewind at normal speed, key 6 to go Forward at normal speed, key 7 to Rewind at 4× speed, and key 9 to go Forward at 4× speed. You can Loop a Selection at normal speed by pressing key 0.

For more options, you can use pairs of keys: press keys 1+2 to Rewind at 1/16× speed, keys 2+3 to go Forward at 1/16× speed, keys 4+5 to Rewind at half-speed, keys 5+6 to go Forward at half-speed, keys 7+8 to Rewind at double-speed, and keys 8+9 to go Forward at double-speed.

You can recall Memory Locations in Shuttle mode by typing Period (.), the Memory Location number, and Period (.) again – *using your computer's numeric keypad*.

Organising the Mix Window

In the Mix window, tracks appear as mixer channel strips with controls for signal routing, input and output assignments, volume, panning, record enable, automation mode, and solo/mute.

From the View menu, you can choose whether to display the Mic Preamps, Instruments, Inserts, Sends, Meters & Faders, Delay Compensation, Track Color, or Comments in the Mix window.

The Group Settings Pop-Up Selector

Directly above the Pan controls at the top of each fader section, there is a pop-up selector that lets you control any Group settings that you have made for the track – see Figure 2.39.

Using this pop-up selector, you can select or hide all the tracks in the group, show only the tracks in the group, delete, duplicate, or modify the group, and see which tracks and attributes are included in the group.

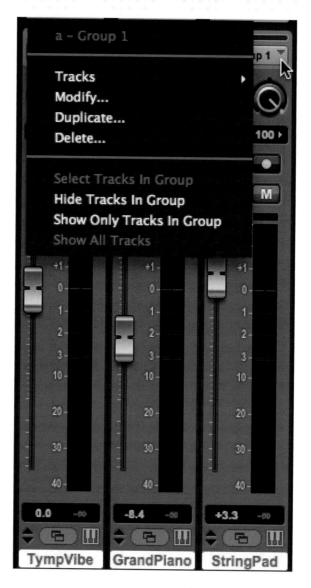

Figure 2.39
Pro Tools Mix window with mouse pointing to the Group settings pop-up selector. The pop-up selector for an adjacent track is open in front of the Mix window.

Inserts

Pro Tools lets you use up to 10 Inserts on each Audio track, Auxiliary Input, Instrument Track, or Master Fader. Each insert can be either a software plug-in or an external hardware device. There are two sets of Inserts, labelled A–E and F–J, so you can conserve space onscreen by only displaying one set at a time.

Sends

You can also use up to 10 Sends on each track. Sends let you route signals across internal buses or to audio interface outputs, so that one plug-in or one external signal processor can be used to process several tracks at once.

There are two sets of sends (labeled A–E and F–J) that can optionally be displayed in the Mix window. You might use the first set to send to effects {such as reverb) that you wish to apply to several tracks and use the second set to send cue mixes to musicians – routing these from your Pro Tools hardware interface to suitable headphone amplifiers.

If you are not using the second set of sends, or the mic preamps, or you don't need to see the delay compensation controls or comments, or whichever, then you should hide these so that they don't distract you from the controls you do want to use and so that the Mix window takes up less space on your computer screen.

I/O Selectors

Track Input and Output Selector pop-ups are located below the Inserts and just above the Automation controls on the channel strips. The Track Input Selector pop-up (see Figure 2.40) lets you choose input sources for Audio tracks, Instrument tracks, and Auxiliary Inputs. Track input can come from your hardware interface, from an internal Pro Tools bus, or from a plug-in.

The Track Output Selector pop-ups let you route the audio from each track to your choice of available outputs or bus paths.

Figure 2.40
Track Input Selector
pop-up.

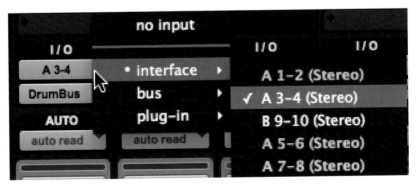

Assigning Multiple Track Outputs and Sends

Pro Tools Audio tracks, Instrument Tracks, and Auxiliary Inputs can have multiple track output and send assignments chosen from the actual paths and resources available on your system (although Master Faders can only be assigned to a single path). Assigning to multiple paths is an efficient way to route an identical mix to other separate outputs, for simultaneous monitor feeds, headphone mixes, or other situations where a parallel mix is needed.

To assign an extra output, hold down the Control key (Mac) or Start key (Windows), open the Output Selector, and select your additional output. A '+' sign is added to the Output Selector legend to remind you that this track has more than one output assigned (see Figure 2.41), and you can add as many additional outputs as are available on your system. If you also hold the Option (Alt) key at the same time as the Control (Start) key, the additional output will be added to all tracks (apart from Master Faders and MIDI tracks, of course).

Figure 2.41
Making multiple assignments for a track's Audio Output Path.

> **NOTE**
> You can use the same procedure to add additional output assignments to track Sends – see Figure 2.42.

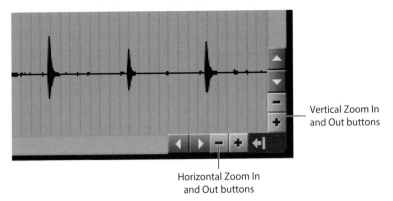

Figure 2.46
Horizontal and vertical zoom buttons

Vertical Zoom In and Out buttons

Horizontal Zoom In and Out buttons

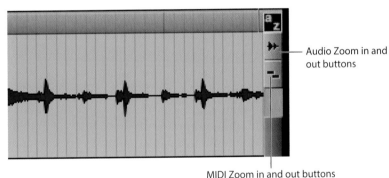

Figure 2.47
Audio and MIDI Zoom In and Out buttons

Audio Zoom in and out buttons

MIDI Zoom in and out buttons

The Tool Buttons

Located by default to the right of the Zoom buttons, you will find six Tool buttons.

Zoomer Tool

The first button is the Zoomer Tool. You can use this to zoom the display either vertically or horizontally. If you click and hold this, you can select the Single Zoom mode. With this selected, you can click in the Edit window to zoom the display one time, then when you let go of the mouse, the Zoomer Tool is de-selected and the Selector Tool becomes active.

Trimmer Tool

The next button is the Trim tool. If you click and hold this, you can select various modes using the pop-up that appears. You can use the standard Trim tool to lengthen or shorten clips or notes. The Time Compression/Expansion

Trim tool (TCE Trim) lets you apply Time Compression/Expansion directly in the Edit window. The Loop Trim tool lets you create or trim looped clips. Pro Tools HD also includes the Scrub Trim Tool. This lets you drag in a track to hear the audio, then trim at a specific location by releasing the mouse button.

Selector Tool

To the right of the Trimmer Tool is the Selector Tool that lets you use the cursor to select areas within the Edit window. With this tool selected, the mouse pointer changes to an insertion cursor that you can use to drag across and select clips in the Edit window or you can simply point and click at a particular location in the Edit window to position the insertion point (and update the Location Indicators) at that location.

> **TIP**
>
> Whenever you are not specifically using one of the other tools, the Selector Tool is the most useful (and safest) tool to leave selected.

Grabber Tool

To the right of the Selector Tool, you will find the Grabber Tool – the one with the 'hand' icon. You can use this to move clips around in the Edit window. You can also use the Grabber to automatically separate an edit selection and move it to another location or another track using its Separation Mode.

The Grabber Tool offers a third mode – the Object mode – that you can enter using the pop-up that appears when you click on the Grabber Tool. This 'Object Grabber' lets you select non-contiguous clips on one or more tracks. Just take a look at Figure 2.48 to see an example of a 'non-contiguous' selection of two clips, all on different tracks. By the way, 'non-contiguous clips' in this context basically means clips that are not next to each other on the same track, or are on different tracks.

Figure 2.48
The Object Grabber Tool being used to select non-contiguous clips in the Edit window.

Smart Tool

The Trimmer, Selector, and Grabber tools can be combined using the Smart Tool to link them together – see Figure 2.49. The Smart Tool button is positioned both above and to either side of the Trimmer, Selector, and Grabber tools. To activate it, simply click in this area above or to either side of these tools. With the Smart Tool activated, then, depending on where you point your mouse in the Edit window, one or other of these tools will become active – saving you having to click on these tools individually when you want to change to a different tool.

Figure 2.49
The Smart Tool.

Scrub Tool

To the right of the Grabber Tool is the Scrub Tool that lets you 'scrub' back and forth over an edit point while you are trying to hear the exact position of a particular sound – rather like moving a tape back and forth across the playback head in conventional tape editing.

Pencil Tool

The sixth tool is the Pencil that you can use to redraw a waveform to repair a pop or click. Alternate Pencil modes are available that constrain drawing to lines, triangle, square, or random shapes.

The Mode Buttons

Running underneath the Tool buttons, there are six more mode buttons.

Zoom Toggle

The leftmost button controls the Zoom Toggle – see Figure 2.50. This feature lets you define and toggle between zoom states in the Edit window, storing and recalling the Vertical Zoom, Horizontal Zoom, Track Height, Track View, and Grid settings. When Zoom Toggle is activated, the Edit window changes to the stored zoom state, and when Zoom Toggle is disabled, the Edit window reverts to the previous zoom state. Any changes made to the view while Zoom Toggle is enabled are also stored in the zoom state.

Figure 2.50
Zoom Toggle button.

> **TIP**
> When Zoom Toggle is enabled, you can cancel it and remain at the same zoom level by pressing Option-Shift-E (Mac) or Alt-Shift-E (Windows).

Tab to Transients

The Tab to Transients button (see Figure 2.51) lets you automatically locate the cursor to the next transient while editing waveforms. With this activated, just press the Tab key on your computer keyboard to jump to the beginning of the next transient (e.g. at the start of a snare hit) in the waveform that you are currently editing.

Figure 2.51
Tab to Transients button.

Mirrored MIDI Editing

The third button from the left (see Figure 2.52) lets you enable or disable Mirrored MIDI Editing. Mirrored MIDI Editing is useful when you edit a clip containing MIDI notes and you want these edits to apply to every MIDI clip of the same name.

Figure 2.52
Mirrored MIDI Editing button.

Automation Follows Edit

Figure 2.53
Automation Follows
Edit button.

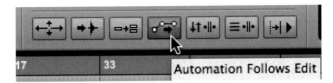

Figure 2.53
Automation Follows
Edit button.

When Automation Follows Edit is enabled, (see Figure 2.53 above), automation events are affected by edits to audio or MIDI notes.

Link Timeline and Edit Selection

The Link Timeline and Edit Selection button (see Figure 2.54) lets you link or unlink Edit and Timeline selections. With the Timeline and Edit selections *unlinked*, you can select different locations in the Timeline at the top of the Edit window and in the clips within the Edit window. With the Timeline and Edit selections *linked*, whenever you make a selection inside the Edit window, the same selection is automatically made in the Timeline at the top of the Edit window.

Figure 2.54
Link Timeline and Edit
Selection button.

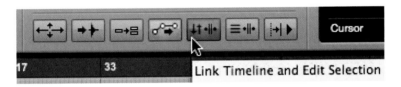

Link Track and Edit Selection

The Link Track and Edit Selection button (see Figure 2.55) also does what it says: with this highlighted, when you select a clip in the Edit window, the track becomes selected as well. If you then select another track, Pro Tools selects the corresponding clip (to the clip selected in the first track) in this other track – because the Track and Edit selection features are linked!

Figure 2.55
Link Track and Edit
Selection button.

Insertion Follows Playback

The Insertion Follows Playback button (see Figure 2.56) lets you enable or disable the 'Timeline Insertion/Play Start Marker Follows Playback' option in the Operation Preferences and also provides a visual indication of whether this option is on. Previously, this option was only available in the Operation Preferences.

Figure 2.56
Insertion Follows Playback button.

When this button is highlighted, the 'Timeline Insertion/Play Start Marker Follows Playback' option is enabled, and both the Timeline Insertion and the Play Start Marker will move to the point in the Timeline where playback stops. When deselected, the Timeline Insertion and Play Start Marker do not follow playback, instead they return to the location where playback began.

> **TIP**
> There is a useful keyboard command that lets you switch this preference on and off: press Control-n (Mac) or Start-n (Windows) or, with the Commands Keyboard Focus enabled, simply press the 'n' key.

Location Indicators

Located centrally at the top of the Edit window you will find the Main and (optional) Sub Location Indicators (or 'Counters') with the Event Edit area to the right of these – see Figure 2.57. The Location Indicators show you where you are in your session in terms of Bars:Beats, Mins:Secs or whatever you have chosen to display here.

The Location Indicators show you where you are in your session in terms of Bars:Beats, Mins:Secs or whatever you have chosen to display here.

The Event Edit Area lets you define selections – by typing the Start and End points, for example – and also serves to display these along with the Length of the selection.

Figure 2.57
Location Indicators
(Counters) and Event
Edit Area.

Edit window MIDI Editing Controls

To allow convenient editing of selected single notes when you are working with MIDI or Instrument tracks, Pro Tools provides various MIDI editing controls (see Figures 2.58 and 2.59) in the Edit window toolbar.

Figure 2.58
MIDI Editing Controls
Area in the Toolbar
above a MIDI track in
the Edit window with
a selected MIDI note,
C2, displayed in the
MIDI Editing Controls
Area together with its
On Velocity of 92 and
Off Velocity of 64.

If you select a MIDI note using the Grabber or Pencil Tools, the MIDI Editing Controls area at the right of the Event Edit area displays the selected MIDI note with its associated MIDI On and Off velocities. You can edit these values by typing new values or by dragging the cursor to change values.

The MIDI socket-with-loudspeaker icon to the right of the MIDI note parameters display can be used to enable or disable the Play MIDI Notes When Editing feature. When this is highlighted in green, MIDI notes will sound when you click on them while editing.

Figure 2.59
MIDI Editing Controls
in the Edit window
Toolbar.

When you are manually inserting notes into a MIDI or Instrument track using the Pencil tool, the pop-up Custom Note Duration selector (see Figure 2.60) lets you choose the default note duration.

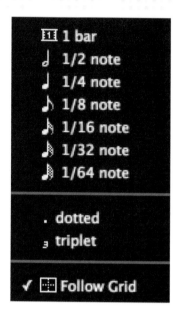

Figure 2.60
Custom Note Duration pop-up selector.

To the right of this, the Default Note On Velocity setting lets you define the default note-on velocity.

Cursor Values Display

Underneath the Counters, there is a display area that tracks the position of the cursor. This area also displays cursor values such as the MIDI note name when the cursor is moved vertically within a MIDI track or the Volume level in decibels when the cursor is moved vertically within a Master track – see Figure 2.61.

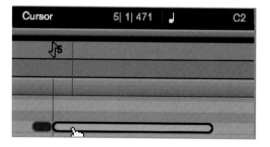

Figure 2.61
Cursor values display.

Edit Window Indicators

Positioned directly below the Event Edit area, there are three small status indicators: Timeline Data Online Status, Session Data Online Status, and Automatic Delay Compensation – see Figure 2.62.

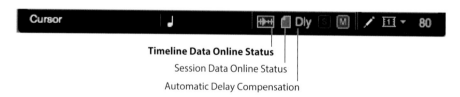

Timeline Data Online Status
Session Data Online Status
Automatic Delay Compensation

Figure 2.62
Edit Window Status Indicators.

If any files that are used in track playlists (i.e. visible in the Timeline: the main part of the Edit window) are offline, being processed, or otherwise unavailable for playback, the Timeline Data Online Status indicator will turn red to warn you.

If any audio or fade files referenced by the session are offline, being processed, or otherwise unavailable for playback, the Session Data Online Status indicator turns red to warn you.

'Dly' indicates that automatic Delay Compensation is active. In PT11, Delay Compensation is always active. In older versions, this was an option that could be enabled or disabled from the Options menu.

Grid and Nudge

Separate displays and controls are provided for the Grid and Nudge features – see Figure 2.63.

Figure 2.63
Grid and Nudge
Displays and controls.

You can click on the small downwards pointing arrows to the right of these displays to open the pop-up selectors that let you change the Grid or Nudge values.

If you are working on music production, you would normally set these values to Bars:Beats, although Samples may be useful for certain types of edits, and Minutes:Seconds is useful if you are not working to Bars:Beats.

The Grid values are used in Grid mode to set the increments by which clips or MIDI notes can be moved forwards or backwards in the Edit window.

The Nudge values are used when 'nudging' clips or notes forwards or backwards in the Edit window. (The word 'nudge' used in this context means to move by a small amount.)

> ### TIP
> To increment or decrement the nudge values using the keyboard, hold the Command and Option keys (Mac) or Control and Alt keys (Windows) and press the plus (+) and minus (−) keys on your QWERTY or numeric keypads.

To nudge a clip, clips, clip groups, MIDI notes, or automation breakpoints, select these using the Time Grabber or Selector tool then press the Plus (+) and Minus (−) keys on the numeric keypad to nudge forwards and backwards in time. Any automation data relating to a selection of clips that you nudge will also be shifted.

> ### NOTE
> Pro Tools can nudge material during playback, so you can nudge continuously in real time to adjust the timing relationship between tracks.

The Rulers

Immediately below the Edit window Toolbar, you will find one or more 'rulers'. See Figure 2.64 below. If you are recording music, the main ruler that you should be using will display Bar:Beats, or possibly Minutes:Seconds. At various times, you will also want to display Markers, Tempo, Meter, Key, and Chords rulers.

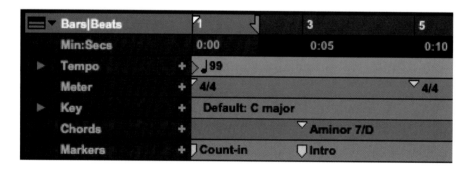

Figure 2.64
A set of Rulers displayed below the Edit window Toolbar.

For certain types of edits, it can be useful to display the Samples ruler. If you are working to picture, you can have up to two Time Code rulers, and for film work you can display Feet+Frames.

You can choose which rulers to display either by selecting these from the Rulers sub-menu in the View menu or by using the pop-up selector at the far left of the main ruler in the Edit window – see Figure 2.65.

Figure 2.65
Rulers pop-up
selector.

The Tempo ruler can be used to insert tempo changes anywhere in your music. Just click on the small '+' sign to open the Tempo Change dialog and insert a tempo change at the current location on the Timeline. The Tempo ruler can also be 'opened' by clicking on the small arrow to the left of its name to reveal the Tempo Editor that allows you to edit tempo 'events' graphically – see Figure 2.66.

The Meter ruler lets you insert time signature markings at appropriate locations along the timeline. Just click on the small '+' sign to open the Meter Change dialog and insert a meter change at the current location on the Timeline.

The Key Signature ruler can be used to insert key changes anywhere in your music. Just click on the small '+' sign to open the Key Change dialog and insert a key change at the current location on the Timeline. Like the Tempo ruler, the Key Signature ruler can also be 'opened' by clicking on the small arrow to the left of its name. This reveals the Key Signature Editor that allows you to edit key changes graphically – see Figure 2.67.

Figure 2.66
'Conductor' Track
Ruler for Tempo with
the Tempo Editor
revealed.

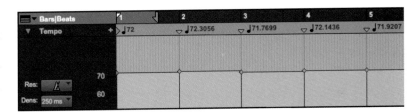

Figure 2.67
Key Signature
Editor.

The Markers ruler lets you insert markers at appropriate locations along the timeline. Just click on the small '+' sign to open the New Memory Location dialog and choose and name a marker and insert this at the current location on the Timeline.

The Chords ruler (see Figure 2.68) lets you add chord symbols to Pro Tools sessions in the Edit window (and in MIDI Editor and the Score Editor windows). Chord symbols in Pro Tools are simply markers that display chord symbols and guitar tablatures – they have no effect on MIDI data.

Figure 2.68

Rulers view showing the Chords ruler above the Markers ruler.

The Chords ruler lets you add, change, move, and delete chord symbols. To add a chord symbol, place the cursor in the Timeline where you want to add this. Click the Plus (+) button in the Chords ruler to open the Chord Change dialog at the current Timeline location. In the Chord Change dialog, you can select the name for the root of the chord, the chord type, the bass note of the chord, and the chord diagram that will be displayed in the Score Editor.

To change a chord symbol, double-click the Chord Symbol marker in the Chord Symbol ruler to open the Chord Change dialog again. If you want to reposition a chord symbol, you can just click and drag the Chord Symbol marker to any new location along the timeline. To delete a chord symbol, either Option-click (Mac) or Alt-click (Windows) on the Chord Symbol marker in the Chords ruler. Alternatively, you can make a selection in the Chords ruler that includes the chord symbols you want to delete and choose Clear from the Edit menu or just press Delete.

TIP

Another way to open the Chord Change, Key Change, Meter Change, Tempo Change, or New Memory Location dialogs that can be convenient is to move the cursor into the ruler while pressing the Control key (Mac) or Start key (Windows). The cursor changes to the Grabber with a '+' and when you click at the location where you want to place the Chord symbol, Key Signature, Time signature, Tempo, or Marker, the appropriate dialog opens.

Edit Window Pop-Up Selectors

Immediately beneath the rulers at the far left of the Edit window you will find two more pop-up selectors. The first of these duplicates the options available in the View menu for the Edit window, allowing you to show or hide Inserts, Sends, and so forth – see Figure 2.69.

Figure 2.69
Edit window View
options pop-up
selector.

Figure 2.69
Edit window View
options pop-up
selector.

To the right of this, you will find the Linearity Display Mode pop-up – see Figure 2.70. This lets you choose between Linear Tick (Bars:Beats) Display and a Linear Sample (absolute) Display.

Figure 2.70
Linearity Display
Mode pop-up.

Track Heights

Pro Tools has seven preset track heights ranging from Micro to Extreme, plus the Fit to window command. MIDI and Instrument tracks also feature a single note view. Micro is the smallest possible Track Height, with six more choices ranging from Mini up to Extreme, plus 'Fit To Window'. 'Fit To Window' fills the entire tracks area of the Edit window (at the current window size) with the selected track: use this setting when you need to focus on a single track for editing.

These track heights can be selected from the Track Options pop-up selector that appears when you click on the small arrow (see Figure 2.71) that can be found at the top left of each set of Track controls in the Edit window.

Figure 2.71
Track Options pop-up selector arrow.

Alternatively, you can press, click, and hold the mouse pointer onto the vertical rule to the left of the waveform display on any audio track (see Figure 2.72) to access the same pop-up selector, which can often be more convenient.

Figure 2.72
Click on the vertical rule to the left of the waveform display to reveal the Track height pop-up selector.

There is a mini-keyboard at the left of each MIDI and Instrument track in the Edit window that lets you select and play notes on that track by clicking with the mouse pointer – see Figure 2.73. To reveal the track height selector, you have to press and hold the Control key while pointing at and clicking on the

Figure 2.73
MIDI Track height pop-up selector – hold the Control key to access this.

mini-keyboard at the left of MIDI or Instrument tracks – otherwise, this action will select and play the notes on the track corresponding to the note on the mini-keyboard that you touched.

Continuously Variable Track Heights

Pro Tools offers yet another way to change the height of a track, resizing in small increments, or continuously if you hold the Command key (Mac) or the Control key (Windows). Simply click and drag the lower boundary of the Track Controls column in the Edit window.

Figure 2.74
Mouse pointer poised, ready to adjust the track height.

When you point your mouse at the bottom line of any Track Controls column (the area at the left of each track in the Edit window containing controls such as Mute, Solo, and so forth), the mouse pointer changes into a thick horizontal line with up and down arrows see Figure 2.74.

If you then press and hold the mouse button and drag upwards or downwards, the track height will change – either in small increments or continuously, depending on whether you are also pressing and holding the Command key (Mac) or the Control key (Windows).

As usual with Pro Tools, there are short cuts provided to let you apply changes to all the tracks, or to a group of tracks that you have selected: To continuously resize all tracks, Option-click (Mac) or Alt-click (Windows) and drag. To continuously resize all selected tracks, Option-Shift-click (Mac) or Alt-Shift-click (Windows) and drag.

TIP

The continuous track height resizing capability makes another couple of useful features possible:

To make all the tracks in which you have made an Edit selection just fit into the Edit window, press and hold the Command and Control keys Arrow (Mac) or the Control and Start keys (Windows) then press either the Up or the Down Arrow key (it doesn't matter which).

To make all the tracks that you currently have selected fit exactly into the Edit window, Option-Shift-click (Mac) or Alt-Shift-click (Windows) onto the Track Height pop-up menu and select Fit to Window.

Zooming Around

Getting used to the different ways to zoom in or out to see your audio and MIDI data displayed appropriately for the tasks you are approaching will take a little while. As usual, Pro Tools provides several ways to do the same things.

Veteran Pro Tools users will be familiar with the group of zoom arrow buttons near the top left of the Edit window that let you zoom audio and MIDI tracks horizontally or vertically. Click the left pointing or right pointing arrows to zoom all the track displays horizontally. Click the leftmost pair of upwards and downwards pointing arrows to zoom the audio waveform vertically and use the rightmost pair of upwards or downwards pointing arrows to zoom the MIDI display vertically.

> **TIP**
> To change the zoom levels continuously, you can simply point, click, and drag on any of these arrow-shaped zoom buttons. Dragging to left or right zooms horizontally in a continuous fashion. Dragging up or down on the Audio Zoom In or Out button continuously zooms all the audio tracks vertically – while preserving any waveform height offsets.

There are five small buttons underneath these zoom buttons that are actually memory locations into which you can store different zoom levels of your choice. Just set the zoom levels you want using the arrow keys (or any other method) then Command-click (Mac) or Control-click (Windows) on any of the five Zoom Presets to store this zoom level.

You can also use the Zoomer tool (the one that looks like a magnifying glass). Just point and click or drag in any track to zoom in, expanding the waveform or MIDI display, or hold the Option (Mac) or Alt (Windows) key while you click or drag to zoom out, contracting the display.

If you hold down the Control key (Mac) or Start key (Windows) while using the Zoomer tool and drag to the left or the right, the display will shrink or stretch horizontally – a bit like a concertina. All tracks are affected equally using this method.

If you drag up or down within any track, the height of the display within the track will zoom vertically. Only the track that you are working with is affected using this method.

> **TIP**
>
> If you double-click the Zoomer tool, the waveform display zooms all the tracks horizontally until everything in your session just fills the Edit window, zooms MIDI to show all notes in a track, and zooms the Tempo Editor to show all the tempo events. This is a very convenient way to reset the display, so that it shows your whole session in the Edit window. This action also resets the waveform display within each track to its default height.

Holding the Shift key as well applies this to all tracks: so if you Control-Shift-drag (Mac) or Start-Shift-drag (Windows) in an audio track with the Zoomer tool selected, this will apply to all the audio tracks (or to all the MIDI or Instrument tracks if you drag in one of these).

You can set all the audio track waveform heights to match the waveform height of the topmost audio track in the Edit window by Command-Shift-clicking (Mac) or Control-Shift-clicking (Windows) on the audio waveform vertical Zoom button.

Pro Tools also offers a keyboard command that zooms the selection to fill the Edit window. Make an Edit selection first, then press Option-F (Mac) or Alt-F (Windows) and *all* the MIDI and Instrument tracks will zoom horizontally until the current edit selection just fills the Edit window. *Selected* tracks will also zoom vertically to make sure that all the MIDI notes within the selection are visible.

Very often, you will want to zoom all the tracks out so that you can see an overview of your session. If you press Option-A (Mac) or Alt-A (Windows), this will zoom all the tracks out all the way horizontally and vertically. MIDI and Instrument tracks will automatically zoom vertically to display all notes. To only zoom horizontally to show the entire session without affecting vertical zoom or scrolling, press Option-Control-A (Mac) or Alt-Start-A (Windows).

To set all audio track waveform heights to match the waveform height of the topmost audio track in the Edit window, Command-Shift-click (Mac) or Control-Shift-click (Windows) the Audio Zoom button. All waveform height offsets will be lost.

To set all MIDI and Instrument track note heights to match the note height of the topmost MIDI or Instrument track in the Edit window, Command-Shift-click (Mac) or Control-Shift-click (Windows) the MIDI Zoom button. All note height offsets will be lost.

> **TIP**
> The only way to become familiar with the way the zoom features work is to keep on trying them out on a regular basis until they become 'second nature' to you – so go back and try these out now!

Zoom Toggle

Using the Zoom Toggle feature, you can make the Edit window behave the way you want it to when you change the Edit selection or Track selection. The Zoom Toggle button is located just underneath the Zoomer tool. Just click on this to activate the Zoom Toggle feature, or, when the Command Focus is active, just press the 'e' key on your computer keyboard.

To configure the way this feature works you need to set various preferences and zoom behaviours in the Zoom Toggle section of the Editing Preferences window, see Figure 2.75, which you can access from the Setup menu. Here, you can set the way that both the Vertical and Horizontal Zoom will switch (i.e. toggle) when you enable the Zoom Toggle feature, and you can also set the way that the Track Height and Track View behave.

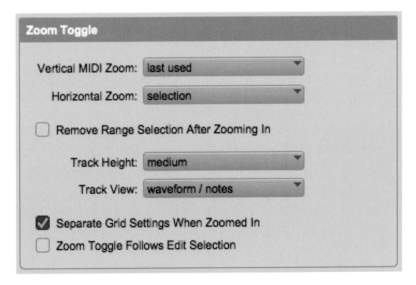

Figure 2.75
Zoom Toggle preferences.

The Vertical Zoom defaults to zooming to the current Edit Selection, with 'Selection' chosen in the Zoom Toggle preferences. If you are editing MIDI notes and you are in Notes view, and if you set the Vertical Zoom Toggle preference to 'Last Used', hitting the Zoom Toggle button zooms vertically to the last stored Zoom Toggle state for MIDI notes.

If you choose 'Selection' as the Horizontal Zoom preference, using the Zoom Toggle feature will cause the Edit window to zoom horizontally to the current selection, which is especially useful for audio editing. If you select 'Last Used' instead, Zoom Toggle makes the Edit window zoom horizontally to the last stored Zoom Toggle state. This option is especially useful for editing MIDI notes.

A check box in the Zoom Toggle preferences lets you 'Remove Range Selection After Zooming In', in which case the current Edit selection collapses and becomes an insertion point when you use the Zoom Toggle. This is useful when you know that you are not going to want to keep any selection you have made previously when you zoom in using the Zoom Toggle feature.

You can also set preferences for how the Track Height behaves when you use the Zoom Toggle. If you choose 'Last Used', using the Zoom Toggle changes all tracks containing an Edit Selection to the last used Track Height. If you choose 'Medium', 'Large', Jumbo or 'Extreme', all tracks containing an Edit Selection change to the Medium, Large, Jumbo, or Extreme Track Height, and if you choose 'Fit To Window', all tracks containing an Edit Selection will change their Track Height to a size that will allow them to fit into the Edit window.

Quite often you will want to change the Track View during a Pro Tools session, so the Zoom Toggle also lets you do this. To make the Zoom Toggle change the Track View for audio tracks to Waveform view and for Instrument and MIDI tracks to Notes view choose 'Waveform/Notes' in the Zoom Toggle preferences. If you want Zoom Toggle to change the Track View to the last used Track View that was stored with Zoom Toggle, choose 'Last Used' and if you don't want any change to the Track View, select the 'No Change' option. With Pro Tools version 7.4 and newer versions, the Track view options also include 'Warp/Notes'.

If you want to keep the grid settings constant when you use Zoom Toggle, tick the option for 'Separate Grid Settings When Zoomed In'. Otherwise, the grid setting stored with Zoom toggle will be recalled when you use the Zoom Toggle feature.

There is also a checkbox for 'Zoom Toggle Follows Edit Selection'. When this is ticked, Zoom Toggle automatically follows the current edit selection. When disabled, changing the edit selection has no effect on the currently toggled-in track.

TIP

If you are editing MIDI, the most sensible Zoom Toggle preferences choices are to set Vertical Zoom to 'Last Used', Horizontal Zoom to 'Last Used', Track View to 'Waveform/Notes', and Track Height to 'Fit to Window' – see Figure 2.76.

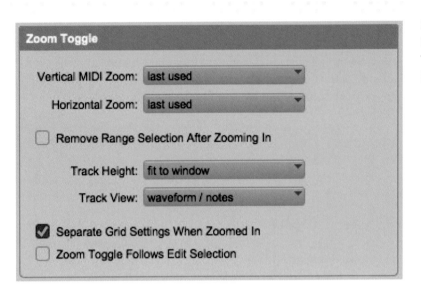

Figure 2.76
Recommended Zoom
Toggle Preferences for
MIDI.

TIP
If you are editing audio, the most sensible Zoom Toggle preferences choices are to set Vertical Zoom to 'Selection', Horizontal Zoom to 'Selection', Track Height to 'Fit to Window', and Track View to 'No Change' – see Figure 2.77.

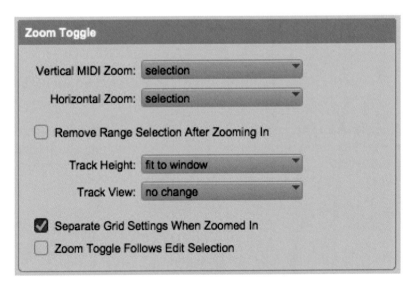

Figure 2.77
Recommended Zoom
Toggle Preferences for
Audio.

NOTE

When you have Zoom Toggle enabled and you select a smaller range of material, the window does not immediately zoom to encompass this new selection – it stays at the same zoom level. However, if you disable Zoom Toggle then enable it again, the window will zoom to encompass the current selection.

Scrolling Windows

There are various scrolling options in Pro Tools. The first of these, 'None', is self-explanatory.

The 'After Playback' option leaves the window where it is when you start playback and scrolls the view to the new position after you stop playback.

The 'Page' scroll option keeps the display stationary until playback reaches the right-most side of the page, then it quickly changes the display to show the next section (i.e. page) of the waveform that will fit within the Edit window – and so forth.

There are two continuous scrolling options that are only available in Pro Tools HD. The first of these, 'Continuous', does exactly what it says – continuously moving the display during playback to keep up with the playback.

The second option, 'Center Playhead', shows the 'playhead' – a vertical blue bar indicating where the playback point is at in the Edit window – and moves the waveforms past this. The Blue Playhead line stays exactly where it is – in the centre of the Edit window – and the items on the timeline move past this – see Figure 2.78.

Figure 2.78
Continuous Scroll
with Center Playhead.

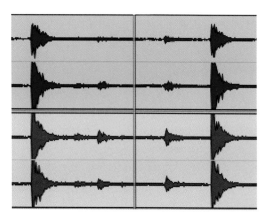

Scrolling Behaviour

With 'Timeline Insertion Follows Playback' enabled in the Operation Preferences dialog, when you press Play, then subsequently press Stop or Pause, then press Play again, Pro Tools 'picks up where it left off'. In other words, when you stop playback, the insertion point 'parks itself' at that point – instead of returning to the position it was at when you commenced playback.

If you are using the 'Continuous' scroll option, the playback position jumps back to the position you started playing from when you pause or stop playback – unless you have enabled the 'Timeline Insertion Follows Playback' option in the Operation Preferences dialog or by clicking the Insertion Follows Playback button in the Edit window.

If you are using the 'Center Playhead' scroll option, the playback position stays wherever it has reached at the moment you pause or stop playback. This happens because this scrolling mode overrides the 'Timeline Insertion Follows Playback' preference, causing Pro Tools to behave as though this preference is enabled – even when it is not enabled.

Note that if 'Timeline Insertion Follows Playback' was disabled before you entered the 'Center Playhead' scrolling mode, it will be still disabled when you exit this mode – which is the way that most users prefer it to behave.

Playback Cursor Locator

The Playback Cursor Locator lets you locate the playback cursor when it is off-screen. For example, if scrolling is not active (set to 'None'), when you stop playing back, the Playback Cursor will be positioned somewhere to the right, off the screen – if it has played past the location currently visible in the Edit window. Also, if you manually scroll the screen way off to the right, perhaps to check something visually, then the Playback Cursor will be positioned somewhere to the left off the screen.

To allow you to quickly navigate to wherever the Playback Cursor is positioned on-screen, you can use the Playback Cursor Locator button – the small blue arrowhead that you can see in Figure 2.79.

A click on the Playback Cursor Locator immediately moves the Edit window's waveform display to the Playback Cursor's current on-screen location – saving you lots of time compared with any other way of finding this location.

Figure 2.79
The Playback Cursor
Locator, the small blue
arrowhead, can be seen
at the top right in the
Main Timebase Ruler.

NOTE
The Playback Cursor Locator button only appears under certain conditions:

It will appear at the *right* edge of the Main Timebase Ruler after the playback cursor moves to any position *after* the location visible in the Edit window.

It will appear at the *left* edge of the Main Timebase Ruler if the playback cursor is located *before* the location visible in the Edit window.

NOTE
The Playback Cursor Locator changes colour from blue to red when any of the tracks are record enabled.

Keyboard Focus

To allow the computer keyboard to be used to issue more commands than there are available keys, Pro Tools provides three types of 'Keyboard Focus': the Commands Keyboard Focus, the Clips List Keyboard Focus, and the Group List Keyboard Focus.

Obviously, only one of these can be active at a time, so when you engage one Keyboard Focus it will disable the one previously engaged.

Depending on which Keyboard Focus is enabled, you can use the keys on your computer's keyboard to select clips in the Clips List, enable or disable Groups, or perform an edit or play command.

> **TIP**
> You can choose the Keyboard Focus by holding Command-Option (Mac) or Control-Alt (Windows) while you press 1 for the Commands, 2 for the Clips List, or 3 for the Group List.

I normally have the Commands Keyboard Focus enabled so that I get fast access to all my favourite keyboard commands.

Commands Keyboard Focus

The Commands Keyboard Focus button is located in the upper right corner of the tracks pane in the Edit window – see Figure 2.80. When you click on this, the letters 'a' and 'z', and the square outline, change to yellow to indicate that it is activated.

Figure 2.80
Commands Keyboard Focus button, 'a–z': located at the top right of the tracks pane in the Edit window.

With the Commands Keyboard Focus activated, a wide range of single key shortcuts for editing and playback become available on your computer's keyboard. To see these commands listed, choose 'Keyboard Shortcuts' from the Pro Tools Help menu.

> **TIP**
> When the Commands Keyboard Focus is disabled, you can still use these keyboard shortcuts by pressing the Control key (Mac) or Start key (Windows) at the same time as the shortcut key.

Clips List Keyboard Focus

When the Clips List Keyboard Focus is selected by clicking the 'a–z' button in the Clips List, (see Figure 2.81), audio clips, MIDI clips, and Clip Groups can be located and selected in the Clips List by typing the first few letters of the clip's name.

Figure 2.81
Enabling the Clips List
Keyboard Focus.

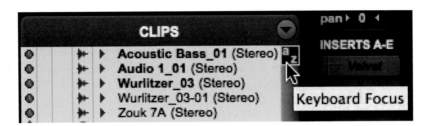

Figure 2.81
Enabling the Clips List
Keyboard Focus.

Group List Keyboard Focus

When the Group List Keyboard Focus is selected by clicking the 'a–z' button in the Edit Groups list (see Figure 2.82), you can enable or disable the Mix and Edit Groups by typing the Group ID letter (a, b, c, etc.) on your computer keyboard when using either the Mix or Edit window.

Figure 2.82
Enabling the Group
List Keyboard Focus.

Soloing Tracks

The default Solo mode is called 'Solo In Place'. In this mode, engaging a Solo button mutes all the other tracks so that the selected track can be auditioned independently. You can enable the solo mode on as many tracks as you like at any time during playback, and if you solo a track that is a member of an active Mix Group this will solo all other tracks that are a member of that active Mix Group as well.

Solo buttons default to Latch mode, in which pressing other solo buttons also solos these tracks without disengaging previously soloed tracks. If you choose X–OR mode, pressing subsequent Solo buttons cancels previously soloed tracks.

In Momentary mode (*Pro Tools HD only*), a track is only in solo while its Solo button is being held down. To temporarily latch more than one solo button in Momentary mode, press and hold the Solo button on the first track that you want to solo, then press additional Solo buttons. These will all remain soloed as long as one Solo button is held.

Solo Safe

In Solo In Place mode, Pro Tools lets you *solo safe* a track to prevent the track from being muted when you solo other tracks. To solo safe a track, Command-click (Mac) or Control-click (Windows) on the track's Solo button. The Solo button changes its appearance in Solo Safe mode, as can be seen in the accompanying screenshot, Figure 2.83.

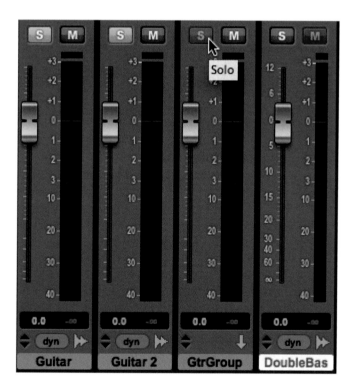

Figure 2.83
Two mono guitar tracks bussed to an Auxiliary Input to form a sub-mix with the Auxiliary Input in Solo Safe mode. Notice the way that the Solo button on the Aux track looks compared with the normal Solo button on the Double Bass track next to it – indicating that the solo-safe is activated for this Aux track.

Solo Safe is useful for tracks such as Auxiliary Inputs that are being used to sub-mix audio tracks, or for effects returns, to allow the audio or effects track to still be heard in the mix when other tracks are soloed. Without this feature, you would have to remember to also solo the appropriate Auxiliary tracks whenever you soloed a track – and this could be very awkward if the Auxiliary track is positioned a long way from the track you are soloing in the mixer.

> **TIP**
> It can also be useful to 'solo safe' MIDI tracks so that their playback is not affected when you solo audio tracks.

Solo Modes

When you are using Pro Tools HD software with Avid HDX or HD Native hardware, you can select two alternative Solo modes, AFL and PFL, from the Options menu – see Figure 2.84. Using these modes, AFL and PFL, the Solo button can be used to route the selected track to a separate AFL or PFL output. You can configure the AFL/PFL output path in the Output page of the I/O Setup dialog.

Figure 2.84
Choosing the Solo mode from the Options menu in Pro Tools HD.

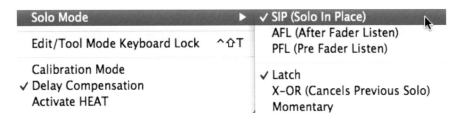

> **NOTE**
> The Solo mode for all soloed tracks can be changed 'on-the-fly' from any Solo mode to either SIP or AFL. Previously soloed tracks will switch their solo behaviour to the new mode. If you try switching the Solo mode for soloed tracks 'on-the-fly' to PFL, all the previously soloed tracks will be taken out of Solo before entering PFL mode to prevent any large boosts in level.

In AFL mode, the Solo button routes the track's *post-fader/post-pan* signal via the AFL/PFL output path. With AFL, the level you hear is dependent on the fader level for that track. In PFL mode, the Solo button routes the track's *pre-fader/pre-pan* signal to the AFL/PFL output path. With PFL, the fader level and pan are ignored, and the level you hear is dependent on the signal's recorded level.

There is also a separate master level setting for AFL and PFL that affects the output of any or all tracks that you solo in AFL or PFL mode. When you are in AFL or PFL mode, you can set this level in the Mix or Edit windows. When you Command-click (Mac) or Control-click (Windows) on a Solo button on any track, a small fader appears (see Figure 2.85) that you can use to adjust the master level for the AFL/PFL Path.

TIP

To set the AFL/PFL Path level to 0 dB, Command-Control-click (Mac) or Control-Start-click (Windows) on any Solo button.

NOTE

You should set the AFL/PFL Output Path to 'None' whenever you are not using this feature. Pro Tools allocates a substantial portion of its available DSP when using AFL/PFL mode, so setting the AFL/PFL Path to 'None' will free up these DSP resources.

Figure 2.85
Setting the AFL/PFL Solo level.

AFL/PFL is optimized for Pro Tools systems using a D-Control or D-Command control surface, where the XMON automatically switches its monitor source between the main output and the AFL/PFL output from Pro Tools. If you are not using a D-Control or D-Command work surface, you will need to choose suitable output paths both for your main monitoring and for your AFL/PFL monitoring in the I/O Setup and manually configure this so that the main Pro Tools output path is muted when you are using the AFL/PFL path. You can

In this chapter

CHAPTER 3

Time operations, tempo operations, and key signatures

Introduction

One of the first things you ought to consider whenever you are working on music in Pro Tools is the tempo. If you are about to record a new piece of music, you will usually want to have the musicians play along to a click to help keep their timing accurate. This will also help to ensure that the music you record lines up with the bar lines in your Pro Tools session so that you can edit using the grid and other editing features. If the musicians don't want to record to a click, there are ways to match the audio you record to the bars and beats in Pro Tools afterwards, but this will inevitably take more time to sort out.

You may also need to create tempo maps so that musicians can play faster or slower in different sections of the music, along with meter changes whenever these are needed. Even short pop songs sometimes have odd numbers of bars or beats or changing meters, and these are very common in music-to-picture. The meter, or time signature, defines the number of beats in a bar and the way they can be sub-divided so you need to be familiar with all these things to get the best results.

It can also be very helpful to establish the correct key signature for your music, especially if you intend to use the Score Editor or key transposition features. And if you want to quickly check whether a guitar or keyboard or bass part or whatever will work better in another key, the Elastic Audio Pitch Transposition feature is the perfect tool for this.

Elastic Audio also lets you treat audio much like MIDI – quantizing and moving notes around more or less at will – using Time Compression and Expansion algorithms in this case. You can use Elastic Audio to quickly beat-match an entire song to the session tempo and Bar|Beat grid. Or you can quantize audio to a groove template so that other audio clips take on the same feel. For instance, you might take the feel from one drum loop that you like and apply this to another. And if you are a fan of Beatles-style recording effects, you can use the Varispeed algorithm to create tape-like effects with speed changes that cause pitch changes.

Beat Detective provides various kinds of analysis of tempo and 'groove' for your audio and MIDI recordings. For example, you can use Beat Detective to establish a 'groove' template that deviates from the standard grid so that you can match new recordings to the 'groove' of a particular drum-machine, sequencer or percussion loop. Beat Detective also lets you analyze the beats in any audio clip, then allows you to separate these so they can be moved around or edited for a variety of purposes. So you can conform beats to the grid in Pro Tools or make the session tempo follow the tempo that your drummer has established.

> **TIP**
> If you are learning Pro Tools for the first time, you should make the effort to get used to the terminology that Avid uses to describe the features of the system so that this all becomes very familiar to you as soon as possible. Then you will find everything less confusing – and when you need to look something up in the manual or remind yourself about how something works, you will be much faster at this.

Timebase and Conductor Rulers

In Pro Tools, you edit and arrange your music along the tracks in the Edit window by placing and moving the audio and MIDI clips along the Timeline that runs from left to right along the window.

Clips and events can be anchored to specific points or locations along this timeline and you can view these locations as time in seconds and minutes, or as SMPTE Time Code locations, or as a bar, beat, and tick locations, and in other formats using the Timebase rulers that can be displayed along the top of the Edit window above the Conductor rulers.

The Timebase rulers can measure time in two different ways: sample-based time (absolute time) as shown in the Minutes:Seconds ruler and tick-based time (relative time) as shown in the Bars:Beats ruler. The relative time that the Bars:Beats ruler represents will change depending on the tempo and meter settings. Absolute time, as displayed in the Minutes:Seconds ruler, can never change, which, of course, is why it is called Absolute time.

Any or all of the Timebase rulers can be displayed at the top of the Edit window in Bars:Beats; or in Minutes:Seconds; or as Time Code, which displays the Time

Scale in SMPTE frames; or Time Code 2, which lets you reference video frame rates that are different from the session Time Code rate; or as Feet + Frames, which is used for film work. You can also display the Time Scale in samples, which can be very useful for precise editing tasks.

The most useful Timebase rulers for music production are Bars:Beats and Minutes:Seconds, so you should de-select the ones that you are not using to remove the potential for confusion and to conserve screen space.

> **TIP**
>
> The Timebase rulers may be used to define *Edit selections* for track material and *Timeline selections* for record and play ranges. For example, using the Selector tool, you can drag along any Timebase ruler to select material across all the tracks in the Edit window. To include the Conductor tracks in the selection, just press Option (Mac) or Alt (Windows) while you drag. Often, you will want to link the Edit and Timeline selections, but sometimes you will want to make different Edit and Timeline selections. You can make or break the link between these using the Link Timeline and Edit Selection button located just under the tools in the Edit window.

The Conductor rulers include the Tempo, Meter, Key Signature, Chords, and Markers rulers. Tempo and meter events entered into these rulers will affect the timing of any tick-based tracks, and also provide the tempo and meter map for the Bar:Beat grid and Click. Pro Tools allows you to edit Tempo events either in the Tempo ruler or using the Tempo Editor that you can open below this – see Figure 3.1.

Figure 3.1
Tempo editor.

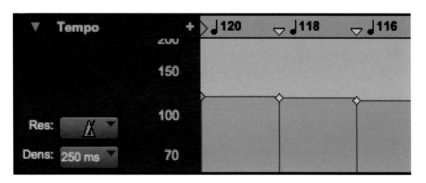

109

You can enter key signatures and make key changes using the Key Signature ruler – see Figure 3.2. This is useful when transposing MIDI notes, for example. When you choose 'Transpose in Key' in the Event Operations Transpose window and enter a particular number of scale steps, Pro Tools works out what the root note of the key should be based on the key signature that applies at this point along the timeline, then transposes any selected MIDI notes by the number of steps that you specify. You can also change the key of audio clips.

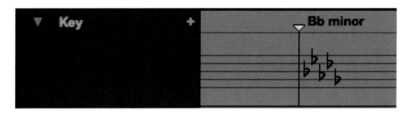

Figure 3.2
Key signature editor.

The Chords ruler (see Figure 3.3) lets you display Chord symbols for musicians to follow and the Markers ruler lets you display Markers to indicate which section of the music is which.

Figure 3.3
Chords and Markers.

Markers are actually Memory locations that can be used for other purposes, but their most basic function, that of marking out the sections of your music, is probably their most important function. It is a good idea to define Markers for any musical session in Pro Tools as early as possible.

Choosing the Marker Reference

When you create or edit Marker (or Selection) Memory Locations in the New or Edit Memory Location dialogs (see Figure 3.4), you can choose whether they have an Absolute (sample based) or Bar:Beat (tick based) reference using the Reference pop-up.

If you choose Bar|Beat, the Memory Location will be tick based and its bar|beat location *will not* change if the tempo is changed. However, its sample location *will* change.

If you choose Absolute, the Memory Location will be sample based and its bar|beat location *will* change if the tempo is changed. But its sample location *will not* change.

So that you can distinguish between the two types of markers, Bar|Beat Reference Markers will appear as yellow chevrons in the Markers ruler, while Absolute Reference Markers will appear as yellow diamonds – see Figure 3.5.

Figure 3.5
Bar|Beat Reference
and Absoluter
Reference Markers.

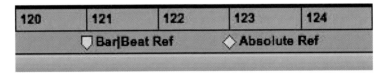

Setting the Time Scales

The Main Time Scale that you should choose for the session will normally be Bars:Beats for music production. This will also be displayed at the Main Counter in the Transport window and at the top of the Edit window. This counter will also be used for the Start, End, and Length values next to the Main Counter, for the Grid and Nudge values in the Edit window, and for any Pre- and Post-roll amounts that you set.

You can also choose which Sub Counter Time Scale to display below each Main Counter. The obvious choice for music production would be Minutes:Seconds so that you always get some feedback about how long clips or complete sessions last.

> **NOTE**
> When the Main Time Scale is set to Bars:Beats, and you are using tempo changes, you should set the Linearity Display Mode to Linear Tick Display. This will keep the Bars:Beats ruler fixed at the selected zoom level, and sample-based rulers such as Minutes:Seconds will scale to fit any tempo changes while bar lengths remain constant.

Meter, Tempo, and Click

If you intend to do anything but the most basic recording and playback of musical audio material you will need to make sure that the bar and beat positions in the audio line up with the bar and beat positions in Pro Tools. It is absolutely essential that the bars correspond correctly if you want to be able to navigate to bar positions and use Grid mode and other features to edit your audio with ease. If you are recording musicians, you can generate a click for the musicians to play along to. If they do this well, then what they play will be in time with the click and will line up with the bar lines in Pro Tools.

Audio that is recorded without listening to a click can still be aligned to the bar and beat boundaries in Pro Tools using Elastic Audio or Beat Detective, or by using the Identify Beat command – but this can take some time. You can also record MIDI with or without a click and either manually add Bar|Beat markers or generate a tempo and meter map using Beat Detective.

Creating a Click Track

It is very easy to create a click for Pro Tools using the Create Click Track command that you will find in the Track menu. When you choose this command, Pro Tools creates a new Auxiliary Input track named 'Click' with the Click plug-in (see Figure 3.6) already inserted and with the click track's Track Height conveniently set to Mini in the Edit window. To hear the click, you also need to highlight the Metronome icon in the Transport window by clicking this, or by pressing the "7" key on your numeric keypad, or by selecting 'Click' in the Options menu.

Figure 3.6
The Click plug-in.

You can use the Click/Countoff Options dialog, (see Figure 3.7), available from the Setup menu, to choose when the click will be active and to set the MIDI notes, velocities, and durations that will play if you are using a virtual instrument or an external MIDI device to produce the click sound. The MIDI parameters in the Click/Countoff Options dialog do not affect the Click plug-in – this is controlled internally by the Pro Tools software. You can also set the number of count-in bars that you prefer and whether you will hear this every time you playback or only when recording.

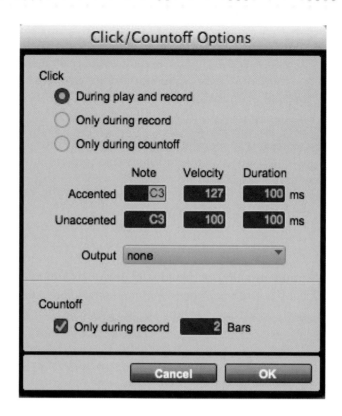

Figure 3.7
Click/Countoff
Options dialog.

Time Signatures and Clicks

When you are setting up a click for musicians to play along too, you may need to change the time signature if the music is not in the default 4/4 meter. This will be very obvious if the music is in 3/4 or 5/4, both of which time signatures are used in popular music. But it may not be quite as obvious if the music is in 6/8 or 12/8.

And if the music is in 6/8, for instance, you may find it better to use a dotted-quarter note value for the click. This way, you would get two clicks in each bar. Otherwise, if you use an eighth note click, this would give you six clicks in each bar which can sometimes be distracting for the musicians.

To set the Meter, double-click on the Current Meter display in the MIDI Controls to open the Change Meter dialog – see Figure 3.8.

Figure 3.8
Double-click the
Current Meter display
in the MIDI Controls
to open the Change
Meter dialog.

In this dialog window, make sure that you are at the correct location in your session by typing in the correct bar, beat and tick values if these are not already correct. If the music uses the same time signature throughout, then you should enter this Meter value at the start of your song or music composition. Obviously, if the meter changes anywhere in the song, you will also need to enter meter changes at these locations.

The Meter Change dialog window – see Figure 3.9 – has a pop-up selector that lets you change the Click 'resolution' – see Figure 3.10.

Figure 3.9
Changing the Click
resolution.

Figure 3.10
Choosing a dotted-1/4
note click resolution.

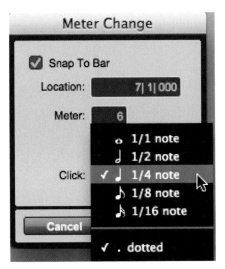

Here you can choose the note sub-division that makes most sense to use. So, for example, you might choose a dotted-1/4 note resolution when you are in 6/8 Meter so that you get just two clicks per bar – see Figure 3.11.

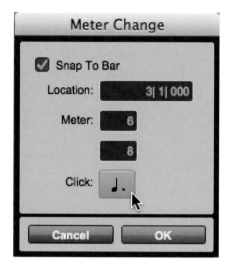

Figure 3.11
Meter set to 6/8 from bar 1, beat 1, with the click resolution set to a dotted-quarter note.

Tempo

The Tempo is one of the most fundamental aspects of music. After all, if there is no tempo, the music goes nowhere! And if the tempo becomes erratic, the music is not 'working' properly. Also, when you are making a recording, getting the tempo to feel 'right' can make all the difference between a successful recording and one that does not inspire people to listen to it. Of course, this can be a subjective matter that not everyone will agree about, but most musicians will perform much better on a particular piece of music if the tempo makes them feel 'comfortable'. And this good feeling will usually communicate itself to the listeners. So, it is very important to pay attention to tempo!

Setting Tempo Manually

You can set a tempo for your session manually using the MIDI tempo controls in the expanded Transport window. When the conductor icon is not selected, you can either type the tempo into the tempo field (see Figure 3.12) or drag the mouse pointer in this field to change the tempo.

Figure 3.12
Setting the tempo using the MIDI tempo controls.

Tapping the Tempo

If you are trying to match the tempo of the session to the tempo of a recording of some music and you have no idea what this should be, just click on the numeric tempo field to highlight it, then tap on the 't' key on your computer keyboard. If you can tap accurately in time with the music, Pro Tools will work out the tempo from your taps and display this for you. Even if you don't get the tempo exactly right, you should at least be able to get this 'into the right ballpark' very quickly using this method.

Using Tempo Events

When you have decided on a tempo for your session, it makes good sense to enable the Conductor track by selecting the Conductor icon in the Transport window and enter the correct tempo for your session into the Tempo Ruler's Song Start Marker.

> **NOTE**
> The small red Song Start Marker located at the beginning of the Tempo ruler defines the initial tempo of the Session and defaults to a tempo of 120 BPM. You can move the Song Start Marker by dragging it to the left or right in the Tempo ruler, but it cannot be deleted.

To edit the default tempo, go to the Tempo Ruler and double-click the red Song Start Marker triangle at the start of the track to open the Tempo Change dialog – see Figure 3.13.

Figure 3.13 Clicking on the Song Start Marker in the Tempo Ruler.

Type the tempo you want to use for your Session in the BPM field and click OK to replace the default 120 BPM value at the start of the Song – see Figure 3.14.

Figure 3.14
Tempo Change
dialog.

Building a Tempo Map

If you know what tempos you want to use for your music, you can build a tempo map for the Conductor track by inserting Tempo events into the Tempo ruler at the locations where you want the tempo to change.

To insert Tempo events, click on the Tempo ruler at the location where you want to insert the tempo event then click the Add Tempo Change button to the left of the Tempo Ruler (see Figure 3.15) to open the Tempo Change dialog.

Figure 3.15
Adding a Tempo
Change to the
Tempo Ruler.

In the Tempo Change dialog, you can type the exact location where you wish to add the new tempo (very useful if you just clicked somewhere nearby this to open the dialog), then type the new tempo in BPM and click OK.

> **TIP**
> To save typing the location if you are close to this, you can tick the 'Snap To Bar' option to place the new tempo event exactly on the first beat of the nearest bar – see Figure 3.16.

Figure 3.16
Tempo Change dialog
with Snap To Bar
option selected.

Editing Tempo Events

If you need to adjust the position of your Tempo events, you can simply drag these back and forth along the timeline. When you drag a Tempo event to a new location, the sample and SMPTE locations for the event are also updated and any neighbouring MIDI events and audio clips on tick-based tracks, along with the ruler, shrink, or expand as necessary to adjust for the new tempo location. However, the BPM value for the dragged tempo event remains constant, as do any other Tempo events in the session.

To delete a tempo event, you can simply Option-click (Mac) or Alt-click (Windows) on the tempo event. To delete several at once, drag to select these using the Selector tool then press the Backspace key or choose Clear from the Edit menu.

If you double-click with the Selector tool in the Tempo ruler, this action will select all the Tempo events. You can then cut, copy, and paste these using the standard computer keyboard commands.

Tempo Operations

You can also apply tempo changes to a time selection using the Tempo Operations dialog that you will find in the Event menu's Tempo sub-menu. There are six Tempo Operations windows to choose from here, including Constant, Linear, Parabolic, S-Curve, Scale, and Stretch.

The first of these, which you will probably use the most often, lets you set a constant tempo at a specified bar position or between two specified bar positions. The next three windows let you create a linear, parabolic or S-curved 'ramp' of tempo changes between a pair of specified bar positions. The final two windows let you scale or stretch existing tempos between specified bar positions. Each window has a basic set of controls and a more advanced set that is revealed when you check the 'Advanced' check box.

For instance, to change the tempo between specific locations, returning to the original tempo afterwards, you would use the Constant Tempo Operations dialog (see Figure 3.17). To see how this works, select, say, Bar 25 through to Bar 29 by dragging the cursor along the rulers, then open the Constant Tempo Operations window from the Event menu. Type the new tempo into the Tempo field and tick the box for 'Preserve tempo after selection'. This causes the tempo to go back to the original value after the selection that you have changed to the new tempo.

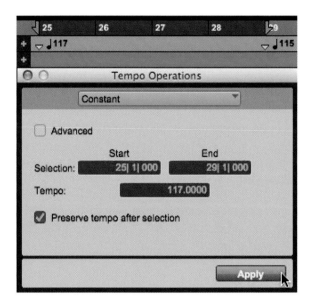

Figure 3.17
Applying a constant Tempo Change to a range of bars.

To ramp up the tempo from one value to another using a series of equal increments over a range of bars, you can use the Linear Tempo Operations dialog (see Figure 3.18) to specify the selection of bars and beats, the start tempo and the end tempo. Using the Advanced settings, you can also specify the resolution and density of the Tempo events that you are creating across this range of bars.

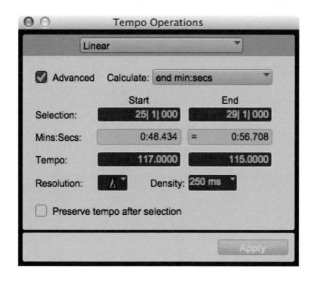

Figure 3.18
Using the Linear
Tempo Operations
dialog to ramp
between tempos over
a range of bars.

The Graphic Tempo Editor

Pro Tools often provides various alternative ways of doing things and tempo editing is no exception! If you click on the small arrowhead at the left of the Tempo Ruler, this will expand downwards to reveal the Tempo Editor. Here you can see tempo information represented graphically.

> **NOTE**
> Tempo events are only visible in the Graphic Tempo Editor when the Conductor is enabled in the Transport window.

You can change the location or value of any Tempo events by dragging these to left or right, up or down, using the Grabber Tool – see Figure 3.19.

Figure 3.19
Using the Trimmer
Tool in the Graphic
Tempo Editor.

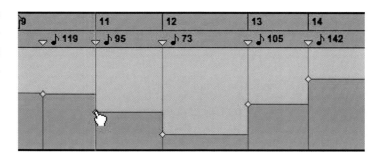

Selected Tempo events can be copied and pasted, nudged, or shifted and you can raise or lower any group of selected Tempo events using the Trimmer tool. First, make a selection that encompasses the Tempo events you want to trim, then just click and drag, up or down, using the Trimmer tool, as in Figure 3.20.

You can also 'draw' in new Tempo events, replacing existing ones, using the Pencil tool. The Line shape lets you draw in a straight line, and with this tool, the Tempo values change in steps according to both the Tempo Edit Density and

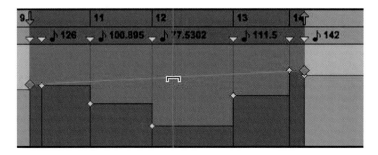

Figure 3.20
Using the Trimmer Tool in the Graphic Tempo Editor.

Resolution. The Free Hand shape lets you draw freely by moving the mouse, producing as a series of steps that depend on the Tempo Edit Density setting. The Parabolic and the S-Curve shapes draw the best possible curve to fit your freehand drawing, again producing a series of steps that depend on the Tempo Edit Density setting. The Triangle, Square, and Random Pencil tool shapes cannot be used to create Tempo events.

There are two pop-up selectors located at the left of the Tempo Editor that let you choose Tempo Resolution and Tempo Edit Density settings for the Tempo events that you draw. Using the 'Res' pop-up selector, see Figure 3.21, you can

Figure 3.21
Tempo Resolution pop-up selector, located above the Tempo Density pop-up.

choose a Tempo Resolution or BPM rate, by selecting a beat note value. The beat note value that you select will normally be based on the meter. So, for example, in 4/4 the beat would be a quarter note and in 6/8 the beat would usually be a dotted-quarter note.

If you select Follow Metronome Click, Tempo events created by drawing with the Pencil tool will automatically mirror the click values set by meter events in the Meter ruler.

Figure 3.22
Tempo Resolution
selector.

NOTE
Although a tempo curve can have different BPM values if there are meter click changes within the selected range, it can be a complicated business to set meter events separately for each Tempo event, so Avid recommends selecting Follow Metronome Click (see Figure 3.22) in most cases.

The Tempo Editor also lets you specify the density of Tempo events created in the Tempo ruler when you draw a tempo curve with the Pencil tool. If you click on the Tempo Edit Density (Dens) selector you can use the pop-up menu (see Figure 3.23) to select a suitable time value either as a note sub-division or in milliseconds.

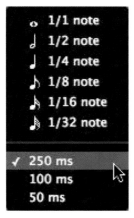

Figure 3.23
Tempo Edit
Density selector.

Changing the Track Timebase

By default, Pro Tools uses a sample-based time for audio, measured in minutes and seconds and tick-based time for MIDI, measured in bars and beats. However, it is possible to change the timebase of audio tracks to be tick based or of MIDI tracks to be sample based.

If the track's height is set to Medium or larger, the Timebase selector button (see Figure 3.24) with its tick-based or sampled-based icon, will be visible among the track controls. When you click on this, a pop-up selector appears (see Figure 3.25), allowing you to change the timebase for the track.

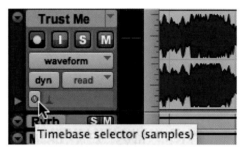

Figure 3.24
Timebase Selector.

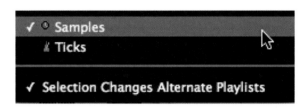

Figure 3.25
Pop-up timebase selector.

If the track's height is set to small or less, click the small arrow to the left of the Track Name (see Figure 3.26) to access the Track Options pop-up menu – see Figure 3.27. Here, you can choose the timebase using the Track Timebase sub-menu.

Figure 3.26
Track Options arrow.

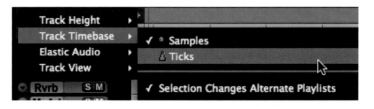

Figure 3.27
Track Options Timebase Sub-menu.

By default, when you change a track's timebase, the change is applied to any alternate playlists that exist on the track. If you only want the change to apply to the active playlist, simply de-select the "Selection Changes Alternate Playlists" option.

> **NOTE**
> When you change the timebase for an audio track that is part of an active group, all the tracks in the group will change to the same timebase.

Sample-based Audio and MIDI

Audio is stored as individual audio samples within audio files on disk. In Pro Tools, clips representing all or parts of these files are used to represent the audio on tracks in the Edit window. Audio tracks are designated as sample based by default. So any particular audible sound is located at a particular sample location on the Timeline. These locations are absolute locations in time measured in samples from the start of the session.

If you change the tempo of the session, the bars and beats will move relative to the absolute (unchanging) locations of sample-based audio clips, but these audio clips and the audio that you hear playing back will not be affected in any way.

> **NOTE**
> If you make a MIDI track sample based, all the MIDI events in the track will have an absolute location on the Timeline – just like the sample-based audio clips – and these will stay fixed to this time location, no matter what tempo or meter changes take place.

Tick-based Audio and MIDI

MIDI (and Instrument) tracks use tick-based timing which ties events to bar|beat locations – not to sample locations. Tick-based locations are relative locations in time, measured in bars, beats, and ticks from the start of the session. So, if you change the tempo of the session, a MIDI clip located at a particular bar|beat location will not move from this bar|beat location – but the location of any sample-based clips in the same session *will* change relative to this bar|beat location.

If you increase the tempo on a tick-based track, the data plays back more quickly – so individual events take place sooner in time relative to any sample-based audio in the session. If you decrease the tempo, events will play back more slowly – so individual events will take place later in time relative to any sample-based audio in the session. Another way of thinking about this is to keep in mind that the bars and beats play back more quickly and last for a shorter time at faster tempos and vice versa.

If you are using Elastic Audio, REX, or ACID audio files, you will want to change the sample-based audio tracks to be tick based, so that the tempo of the audio will automatically change when you change the tempo of the session.

> **NOTE**
> Pro Tools Elastic Audio uses exceptionally high-quality transient detection algorithms, beat and tempo analysis, and Time Compression and Expansion (TCE) processing algorithms. Tick-based Elastic Audio tracks actually change the location of samples according to changes in tempo because the audio stretches or compresses to match these tempo changes.

Keep in mind that tempo changes will only affect the locations of the start points (or the sync points if these are set differently) of any standard (i.e. non-Elastic Audio) audio clips in tick-based audio tracks. You can slice a drum loop into beats in Pro Tools, make the track tick based, and it will play each slice back earlier or later according to the tempo – in a similar way to REX files.

> **NOTE**
> In tick-based audio tracks, the location of an audio clip is determined by the clip's start point, unless the clip contains a sync point. If the clip contains a sync point, the sync point determines where the audio clip is fixed to the grid.

Time Scales and Tick Resolution

When you are working on music in Pro Tools, you will normally set the Main Time Scale to Bars:Beats with the Sub Counter Time Scale set to Mins:Secs. Because you are working with music, it will make more sense to use bar, beat and tick values for a number of operations, including placing and spotting clips; setting lengths for clips or MIDI notes; locating and setting play and record ranges (including pre- and post-roll); specifying settings in the Quantize and Change Duration pages of the Event Operations window; and setting the Grid and Nudge values. You would not normally want to set these using samples, although this may be useful at times.

With the tick-based Time Scale, the bars and beats are sub-divided into clock 'ticks' with each quarter note represented by 960 ticks. This is often quoted as a tick resolution of 960 PPQN (pulses per quarter note), and all MIDI sequencers have a similar specification. Some MIDI sequencers have the same resolution as Pro Tools, while others offer less or more PPQN, and some allow you to select the resolution that you feel is appropriate.

> **NOTE**
> In Pro Tools, the actual internal MIDI resolution is 960,000 PPQN, with 960 PPQN used as the display resolution so that the numbers are more manageable.

The Linearity Display Mode

You can choose to view the Timeline using either a Linear Sample scale or a Linear Tick scale that you can choose using the Linearity Display Mode selector – see Figure 3.28. The Linear Sample scale uses an absolute timebase while the Linear Tick scale uses a relative timebase, so MIDI and Instrument tracks, audio tracks, and Tempo curves will look and behave quite differently depending on which of these you choose.

Figure 3.28
Linearity Display
Mode Selector.

127

When you are working primarily with sample-based material, which is usually the case when recording or mixing audio, you should use the Linear Sample Display mode – especially if you intend to align the tempo and meter map to sample-based events. In Linear Sample Display mode, because the Timeline display is sample based, the locations of tick-based events such as bars and beats will shift whenever you change tempo. So, for example, the locations of the bars and beats will move to different time positions – as you will see if you display minutes and seconds on the sample-based Timeline.

Using the Linear Sample display, the Min:Secs ruler shows each second occupying the same distance along the timeline. Looking at the screenshots shown here, you can see that the number of seconds displayed stays the same (with no change made to the zoom level), while the corresponding number of beats displayed depends on the tempo. At slower tempos, fewer bars and beats play back in a given time, while at faster tempos, more bars and beats play back in that same time interval. Another way of looking at this is to notice that, as the tempo increases, the length of the bars along the timeline gets shorter – see Figures 3.29 and 3.30.

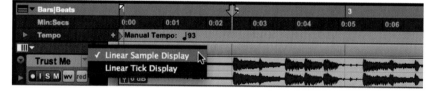

Figure 3.29

Each beat at 93 beats per minute occupies around 2.6 seconds. So in the 7 seconds shown in this example, about 3 complete bars will play back.

Figure 3.30

Each beat at 234 beats per minute occupies about 1 second. So with the Linear Sample display selected, you will see more bars playing back within the same number of seconds along the timeline – 7 bars playing back in the 7 seconds displayed in this example.

When you are working with tick-based material, such as MIDI notes or Elastic Audio events, you may find it more useful to switch to the Linear Tick Display mode – see Figure 3.31. When you make tempo changes in Linear Tick Display mode, the bars and beats will remain fixed in the Timeline while sample-based events will move to a different bar and beat locations along the Timeline, as can be seen in Figure 3.32.

Figure 3.31
At 93 BPM, 7 bars are displayed here in Linear Tick mode, while the time elapsed shows about 17 seconds.

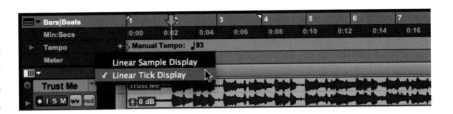

Figure 3.32
At 234 BPM, 7 bars are still displayed with the timeline display in Linear Tick mode (with no change made to the zoom level), but the time elapsed shows just 7 seconds.

In Linear Tick Display mode, all the bars are evenly spaced, whatever the tempo. So, for example, sample-based audio clips will expand along the timeline as you increase the tempo – although they will still play back the audio that they contain in the same length of time. Tick-based material, such as MIDI notes or Elastic Audio material on tick-based tracks, will play back at whatever tempo you set the session to.

> **TIP**
> You should use the Linear Tick Display when drawing tempo changes, for example, because using the Linear Sample Display causes Bar|Beat-based material to move, making it awkward to work on tick-based material such as Tempo events.

129

> **NOTE**
> When the Tempo Edit Density setting (in the Tempo Editor) and the Linearity Display Mode setting are both set to either a Bars|Beats Time Scale or to an absolute Time Scale, Tempo edits appear evenly spaced. With either one set to an absolute Time Scale and the other set to a Bars|Beats Time Scale, the number of Tempo edits will appear to increase or decrease over time (against absolute time).

Tempo Events and Bar|Beat Markers

Tempo events are for use with tick-based tracks, while Bar|Beat Markers are intended for use with sample-based tracks. They cannot be mixed, so if a session contains Tempo events and you insert Bar|Beat Markers, existing Tempo events are converted to Bar|Beat Markers (and vice versa).

> **NOTE**
> Tempo events are displayed as small green triangles in the Tempo ruler. Bar Beat Markers look similar to Tempo events, but are blue in colour.

Tempo events can be manually converted to Bar|Beat Markers and vice versa using the Tempo Ruler pop-up menu. When you hold down the Command (Mac) or Control (Windows) key on your computer keyboard and click the 'Add Tempo Change' button, this opens the Tempo ruler pop-up menu where you can select which of these you wish to use – see Figures 3.33 and 3.34.

Figure 3.33
Clicking while holding the Command (mac) or Control (Windows) key to convert Tempo events to Bar|Beat Markers.

Figure 3.34

Using the Tempo
Ruler pop-up menu to
change Tempo events
to Bar|Beat Markers.

NOTE

If you build a tempo map in a MIDI sequencer, such as Digital Performer, Logic or Cubase, then transfer this to Pro Tools using a MIDI file, this tempo map will appear as Tempo events in Pro Tools.

If you are working primarily with MIDI instruments in Pro Tools, it makes sense to use Tempo events for any tempo changes. On the other hand, if you are working primarily with audio, and you want to map out where the barlines and beats fall within audio material that was not recorded to a click, it makes better sense to use Bar|Beat markers.

Building a Tempo Map by Inserting Bar|Beat Markers

You can insert individual Bar|Beat Markers anywhere in the timeline by setting an Edit insertion point (instead of making a selection), then using the Identify Beat command. This ability to identify each individual beat is a particularly useful technique to use when you just need to make a few corrections to tighten up the correspondence between the Pro Tool bars and beats and the audio that you are working with. If the tempo gets a little faster or slower, even within a bar, you can easily insert a new Bar|Beat Marker to make sure that Pro Tools keeps in step with your audio.

In the example shown in Figure 3.35, this drumbeat was played a little late at bar 7, beat 1, clock tick 023.

To move the beat in Pro Tools to correspond with the actual drum beat, make an Edit insertion point using the Selector tool at the beginning of the beat, then use the Identify Beat command to open the Add Bar|Beat Marker dialog. Use this dialog to tell Pro Tools that this Edit insertion point is where this beat is supposed to be by typing the correct bar beat and clock tick location.

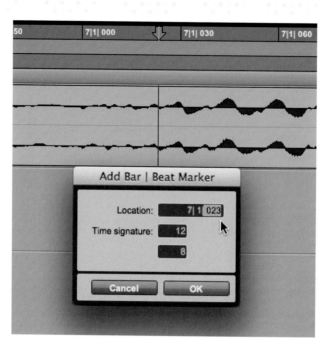

Figure 3.35
Inserting a Bar|Beat Marker at an Edit insertion point.

When you OK the dialog, Pro Tools inserts a beat marker at this point, along with a tempo event that adjusts the Bar|Beat location to correspond with this point – see Figure 3.36.

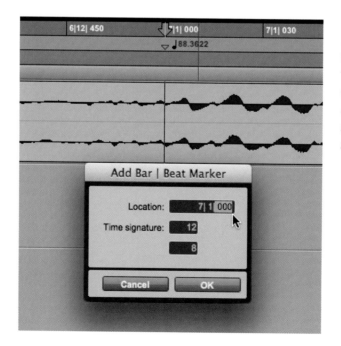

Figure 3.36
Bar|Beat Marker inserted into the Tempo Ruler to make this beat in Pro tools correspond with the actual beat in the audio.

TIP

To delete a Bar|Beat Marker, just hold down the Option (Mac) or Alt (Windows) key, move the cursor over the Bar|Beat Marker, and click to remove it.

When you have your Bar|Beat markers in place, you may need to finely adjust the locations to position these even more accurately – dragging each marker, as necessary, to align with the associated beat within the audio.

NOTE

When you drag Bar|Beat markers back and forth along the timeline to re-position their locations, this results in neighbouring MIDI data being adjusted to align with the new tempo map.

Re-positioning Bar|Beat Markers

Because Bar|Beat Markers are sample based, they behave differently from tick-based Tempo events when you drag them in the Tempo ruler. When you drag a Bar|Beat Marker, its bar and beat location is dragged *with* the Bar|Beat Marker. This location does not change.

So, for example, if the Bar|Beat Marker was originally placed at 3|1|000, it will still be at this location (unless you specifically change this). However, its BPM value is recalculated along with the BPM value of the Bar|Beat Marker to its immediate left. Also, its sample and SMPTE locations will be recalculated according to the new tempo of the Bar|Beat Marker. Bar|Beat Markers to the right of the dragged marker remain unchanged. Neighbouring MIDI events, along with the Bars|Beats ruler, shrink or expand as necessary to adjust to the new tempo.

If you want to change the bar and beat location of a Bar|Beat marker, you must use the Edit Bar|Beat dialog instead. If you double-click on a Bar|Beat Marker, this

opens the Edit Bar|Beat dialog – see Figure 3.37 – where you can make changes
to its bar and beat location and also change the time signature if you wish.

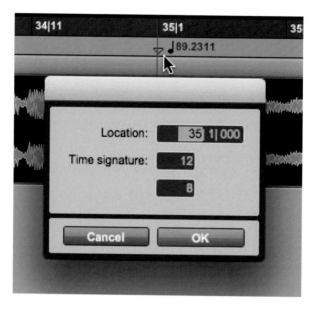

Figure 3.37
Edit Bar|Beat
Marker Dialog.

Time Operations

The Time Operations window provides four important functions controlled on
separate pages. You can use this to change the meter. Insert or cut time, or
move the Song Start position.

Setting the Meter

Pro Tools defaults to 4/4 meter, which is the time signature of most popular
music. If you want to work in other meters such as 3/4 (waltz time) or 5/4 (like
Dave Brubeck's jazz hit "Take Five") or with the changing meters that are used
in various forms of music such as classical and orchestral film scores, then you
will normally need to set these before you start recording.

To open the Meter Change dialog (Figure 3.38), you can either click on the +
sign in the Meter ruler or double-click on the displayed time signature in the
MIDI control section of the Transport window.

Figure 3.38
Meter Change
window.

Alternatively, you can use the Change Meter dialog (Figure 3.39) that you can select from the Event menu's Time Operations sub-menu.

Figure 3.39
Time Operations
Change Meter dialog.

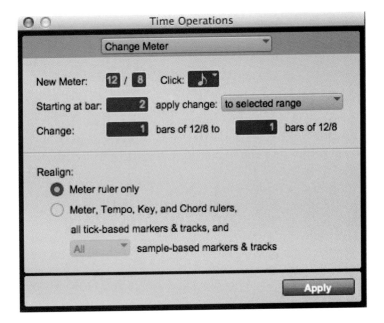

One situation where you will need to map out tempo or meter changes is when you have recorded a musician playing without a click, perhaps to capture an improvised performance. This could involve meter changes where the musician plays a bar of 3/4 or a bar of 5/4 by just 'feeling' out how many beats to play in a particular section. Sometimes this is done deliberately, sometimes not, but, whatever the reason, you can always use the Tempo and Meter rulers to insert changes that follow the performance, as in Figure 3.40.

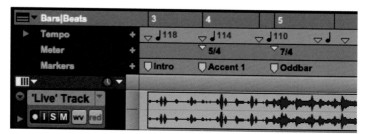

Figure 3.40
Tempo, Meter, and Markers rulers constructed to follow an improvised performance.

Insert or Cut Time

When you are working on your musical arrangements, it is quite likely that you will need to insert some bars or remove some bars at some time or other.

You could just move everything in your session forwards or backwards by the correct number of bars, but this can be quite awkward and is not what you usually want to do. Pro Tools has the features that you need to do this conveniently and properly – adding or removing 'time' to or from the timeline.

The 'Time Operations: Insert Time' dialog (Figure 3.41) lets you insert time from any particular start location and lets you specify the amount of time you wish to insert in terms of bars, beats, and clocks. It also moves any tempo markers forwards from the insertion point.

The 'Time Operations: Cut Time' dialog (Figure 3.42) lets you remove time, starting from any particular start location, and lets you specify the amount of time you wish to remove in terms of bars, beats, and clocks. This also moves any tempo markers backwards from the insertion point.

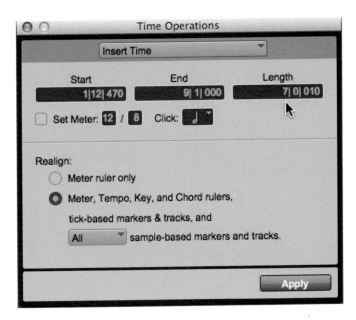

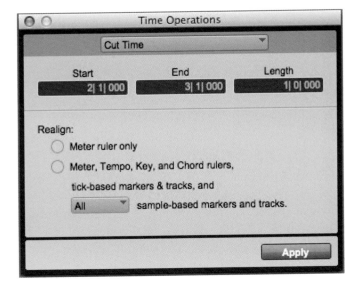

Move Song Start

Sometimes you may need to move the actual start point of the song, so Pro Tools provides the 'Time Operations: Move Start' dialog (Figure 3.43) to let you specify the bar, beat, and clock location that you want to move it to.

Figure 3.43
Time Operations:
Move Song Start
dialog.

Normally, all tick-based markers and tracks will move, along with any meter and Tempo events. You can choose whether or not to move sample-based markers and track.

Key Signatures

Pro Tools provides a Key Signature ruler for adding key signatures to Pro Tools sessions. Key signatures are an important part of any musical composition that will be displayed as notation for musicians to read. Now that Pro Tools has a Score Editor and separate MIDI Editor windows, it is becoming much more likely that you will need to enter Key Signatures as necessary into the Key Signature ruler for each Pro Tools session. The default key signature is C major, but it is highly unlikely that every piece of music you work on will be in this key!

Key signatures can be imported and exported with MIDI data. This is especially useful when exporting MIDI sequences for use in notation programs like Sibelius and can save you lots of time if you have an existing MIDI sequence to import to help you get started with your music production.

All MIDI and Instrument tracks in Pro Tools now have an option called Pitched. When this is enabled, these tracks will automatically be transposed or conformed to any key signature changes – although this should be disabled on drum tracks (which should not normally be transposed). You can also make quick transpositions using the Real-Time Properties for any track or clip.

The Key Signature Ruler

The Key Signature ruler (see Figure 3.44) lets you add, edit, and delete key signatures. You can use key signatures to indicate key and key changes in your Pro Tools session. Key signatures can also be used for various other functions, such as transposing in key or constraining pitches to the specified key.

Figure 3.44
Key Signature ruler.

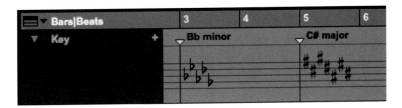

To view the Key Signature ruler, select this using the Rulers sub-menu in the View menu or from the pop-up View selector near the top left of the Edit window. If you click on the small Show/Hide triangle at the left of the Key ruler, this will reveal the Key Signature Staff.

To the right of the ruler's name in the Edit window there is a small '+' button that can be used to add a key signature at the current timeline position. Alternatively, you can choose 'Add Key Change' from the Event menu. Using the Key Change dialog that appears, you can select the mode (major or minor), the key, its location and range, and how you want it to affect pitched tracks. You can also edit an existing Key Signature by double-clicking its marker in the Key Signature ruler to open the Key Change dialog.

To delete a key signature, simply Option-click (Mac) or Alt-click (Windows) the Key Signature marker in the Key Signature ruler. Another way to do this is to make a selection in the Key Signature ruler that includes the key signature you want to delete and choose Clear from the Edit menu or press Delete.

The Key Change Dialog

Using the Key Change dialog (see Figure 3.45) that opens whenever you add (or double-click to edit) a Key Signature marker, you can choose whether the key is major or minor, select from any of the flat keys or sharp keys, specify a range of bars throughout which the key change will apply, and choose whether or not existing MIDI notes on pitched tracks will be transposed or constrained to the key.

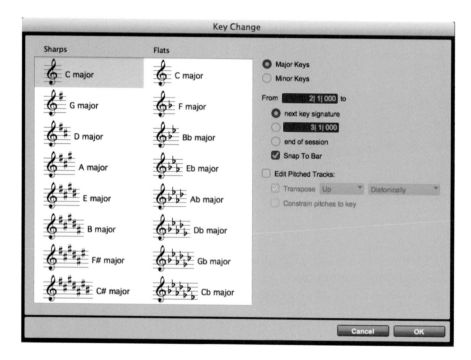

Figure 3.45
Key Change dialog.

You can view the sharp keys in the left-hand column and the flat keys in the right-hand column. To the left of these columns, a pair of 'radio' buttons lets you select major or minor keys (from the natural minor mode). To the right of the key change columns, there are various options. You can specify the Bar:Beat locations at which and until which the key change will be applied (to the next Key Signature marker, to a selection, or to the end of the session). An option is also available to 'snap' the Key Signature marker to the bar line, which can sometimes be useful.

Enabling the 'Edit Pitched Tracks' option lets you transpose existing MIDI notes on Pitched tracks up or down, either diatonically or chromatically based on the key change – or you can constrain pitches to the new key.

For example, if the session is in C major and you want to change the key to D minor, you would select 'Edit Pitched Tracks', 'Transpose', 'Up', and 'Diatonically'. When you OK the dialog, this will transpose notes on Pitched tracks up a whole step and then lower the third, sixth, and seventh scale degrees by a half step to make the key minor. Specifically, the notes C, D, and E would be transposed to D, E, and F natural. If you transposed these notes chromatically instead of diatonically, E would be transposed to F-sharp instead.

So, diatonic transposition alters notes to make them conform to the type of key change that you want, whether major to minor, or minor to major. Any chromatic note alterations that exist in the material being transposed, such as a flattened seventh, are left unaltered by diatonic transposition. So, for example, if you change key from C major to E major and there is a Bb note among those you have selected for transposition, the B-flat is transposed to D-natural – which is the flattened seventh note in the scale of E major.

The 'Constrain Pitches To Key' option lets you constrain pitches to the notes of the new key. When you use this option, any diatonic pitches of the old key on Pitched tracks that are not in the new key are individually transposed to the nearest diatonic pitch of the new key, while notes that are in the new key are left untouched.

For example, when changing from C major to D major, the sequence of the notes C, D, E, F, G will become C-sharp, D, E, F-sharp, G. The way this works is that C gets transposed to C-sharp which is in the scale of D major; D is not transposed because it is already a valid note in the scale of D major; E is not transposed because it is already a valid note in the scale of D major; F gets transposed to F-sharp which is in the scale of D major; and G is not transposed because it is already a valid note in the scale of D major.

Constraining pitches to key also constrain any chromatic pitches to the new diatonic scale. For example, going from C major to D major, the sequence of notes C, D, D-sharp, E will be changed to C-sharp, D, D, E. The way this works is that C gets transposed to C-sharp which is in the scale of D major; D is not transposed because it is already a valid note in the scale of D major; D-sharp is moved to D to keep it within the scale of D major; and E is not transposed because it is already a valid note in the scale of D major,

Diatonic versus Chromatic

This is most simply explained by discussing the notes on the piano keyboard: When you play the black and white keys of a piano one after the other, this is called the chromatic scale and always consists of a series of semitone intervals.

A diatonic scale is what you hear when you play the white keys of the piano one after the other. If you start on C, you get a major scale, also known as the Ionian mode. Starting on A, you get the natural minor scale (the descending form of the melodic minor scale), also known as the Aeolian mode. If you start on any other note, you get one of the other 'church modes', the Dorian starting on D, the Phrygian starting on E, the Lydian starting on F, the Mixolydian starting on G, or the Locrian starting on B. You can also play a major scale starting on any black key and derive diatonic scales from this in a similar way.

Diatonic Intervals are understood to be the intervals between pairs of notes taken from the same diatonic scale. Any other intervals are called chromatic intervals. Diatonic chords are understood to be those that are built using only notes from the same diatonic scale; all other chords are considered to be chromatic. The meanings of the terms diatonic note and chromatic note vary according to the meaning of the term 'diatonic scale', which is defined slightly differently by different authorities. Generally – not universally – a note is understood as diatonic in a context if it belongs to the diatonic scale that is used in that context; otherwise, it is chromatic.

Pitched Tracks

MIDI and Instrument tracks are regarded as "pitched" tracks, because MIDI note data specifies the pitch of the notes. Normally, you will want these tracks to be affected by transpositions due to key changes. But if these tracks are controlling drum machines or samplers, the opposite is true: you normally will not want these tracks to be affected by key changes because it would mess up the key mappings that cause particular drums or samples to be played.

Pro Tools allows you to override the default "pitched" behaviour on individual tracks using the track's Playlist selector – see Figure 3.46. Simply de-select the

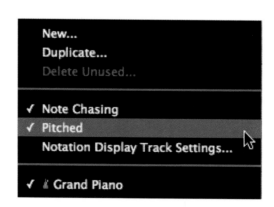

Figure 3.46
Using the Playlist selector to de-select the Pitched option.

Pitched option (so that it has no tick mark next to it in the Playlist selector) for specific MIDI and Instrument tracks – and these won't be affected by key changes.

Diatonic Transposition

Various diatonic transposition options are available from the 'Event Operations: Transpose' dialog (see Figure 3.47) that you can access using the Event Operations sub-menu in the Event menu. The Transpose Event Operations dialog allows you to transpose by octaves and semitones, or from one specified pitch to another, or to transpose all the notes you have selected to a specified pitch. It will also allow you to transpose in key, up or down, by scale steps. For example, if you have a sequence of notes that you want to double in thirds, you can simply copy this to a new track, select the MIDI notes on the new track, and transpose them up by two scale steps. In C major, for example, the sequence C, D, E would be transposed to E, F, G.

Figure 3.47
Event Operations:
Transpose dialog.

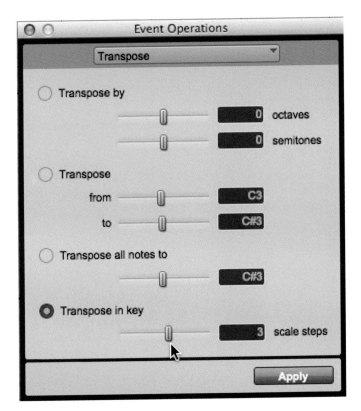

You can also use the track-based MIDI Real-Time Properties (see Figure 3.48) or the MIDI Real-Time Properties window (see Figure 3.49) from the Event menu to transpose up or down by octaves and semitones, to a named pitch, or to a key.

Figure 3.48
Track-based MIDI
Real-Time Properties
Transposition features.

Figure 3.49
MIDI Real-Time
Properties window
Transposition features.

Both of these have a Transpose Mode pop-up selector that lets you choose the type of transposition that you want. The default is to transpose by octave and/ or semitone. You can also transpose to a specified pitch or in key – see Figures 3.50 and 3.51.

Figure 3.50
Transposing to a
specified pitch.

Figure 3.57
Changing the Track
Timebase to Ticks.

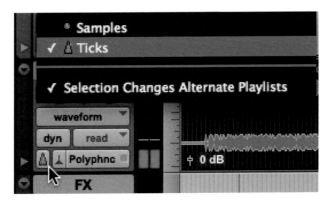

TIP

To change the tempo manually, first make sure that the Conductor icon is de-selected in the Transport window. Use the mouse to point, click and hold, then drag up or down on the tempo field in the Transport window – see Figure 3.58.

Figure 3.58
Changing the tempo
of an Elastic Audio-
enabled track.

NOTE

Although the tempo of the audio will change when you change the session tempo, the audio will not necessarily match the tempo of the session (unless the tempo of the audio is constant and you have originally set the tempo of the session to match this).

TIP

To match a recording with changing tempos to the session tempo, you need to use Elastic Audio's advanced Warp matching features.

Conform to Tempo Command

A very quick way to make the tempo of an audio clip conform to a fixed session tempo is to use the Conform to Tempo command. Assuming that a tempo is already set in Pro Tools, all you need to do is to enable one of Elastic Audio's analysis modes on the track containing the audio clip.

When you select an audio clip within the track, the Conform to Tempo command becomes available in the Clip menu. If you go ahead and choose the Conform to Tempo command, this analyzes the selected clip to determine

> **NOTE**
> Conform to Tempo can only be applied to clips and cannot be applied to clip groups. To conform clip groups to the tempo, you must first ungroup the clip group, then apply Conform to Tempo to the underlying clips, and then regroup those clips – see Figure 3.58.

its tempo and duration in bars and beats. If the tempo of the clip is different from the tempo of the session, Elastic Audio processing is then automatically applied to match the tempo of the clip to the tempo of the session.

Beat Detective

An alternative way of conforming audio to the tempo of a Pro Tools session is to use Beat Detective. Unlike Elastic Audio, Beat Detective does not use time compression/expansion. This has the advantage that it does not affect the quality of the audio that has been conformed. It may introduce audible glitches where clips have been separated, but an Edit Smoothing capability is provided to alleviate this problem.

For example, I used Beat Detective to split the beats in a piano and drums loop into separate clips that I was then able to line up with the bars and beats grid in Pro Tools.

Take a look at the accompanying screens to see how this works. I set the tempo to somewhere close to the tempo of the clip by trial and error – guessing that it was about 100 BPM. Then I selected the clip, looped the playback, and trimmed the start and end points to make sure that this was exactly one bar long. At this point, I could see that the main beats did not line up with the barlines in Pro Tools exactly – see Figure 3.59.

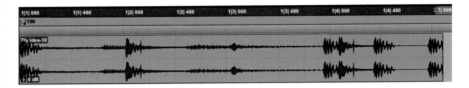

Figure 3.59
A one bar clip
selection. Notice that
the main beats do
not fall exactly
on the barlines.

To sort this out, I opened Beat Detective by holding the Command key (Control key for Windows) and pressing 8 on my numeric keypad.

Using Beat Detective's Clip Separation feature, I captured my clip selection, checked that the start and end points were correctly shown as Bar 1|Beat 1 and Bar 2|Beat 1, made sure that the time signature was correctly shown as 4/4, then chose the appropriate beat sub-divisions (1/4 notes in this case). With this all prepared, I clicked Analyze then adjusted the Sensitivity slider to find the beats within the selection – see Figure 3.60.

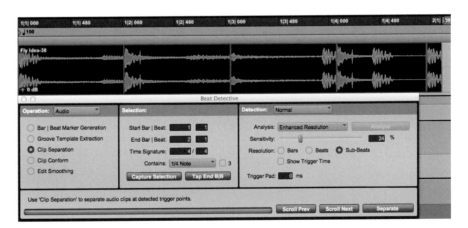

Figure 3.60
Beat Analysis using
Beat Detective.

I noticed that one extra beat was incorrectly identified just before the end of the bar, so I Option-clicked (Alt-click for Windows) on the beat marker using the Grabber tool to remove this before clicking the Separate button in Beat Detective to split the selection into four clips, each containing one beat of the music – see Figure 3.61.

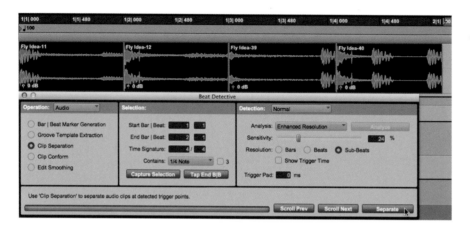

Figure 3.61
Clip Separation using Beat Detective.

The next step in the process is to use the 'radio' buttons at the left of the Beat Detective window to switch to Clip Conform (see Figure 3.62), hit the Conform button and then watch the beats/clips move to line up with the barlines – just like magic!

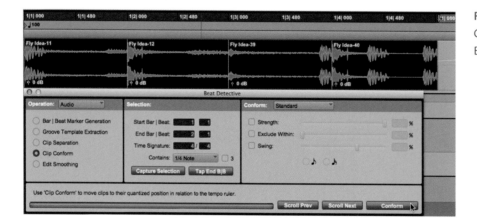

Figure 3.62
Clip Conform using Beat Detective.

I decided that this particular groove sounded better at 96 BPM, so I changed the tempo and conformed the clips to this. This time, because the tempo was slightly lower than the tempo of the original audio, gaps opened up between the clips – see Figure 3.63.

Figure 3.63
Conforming the clips
to 96 BPM leaves
gaps.

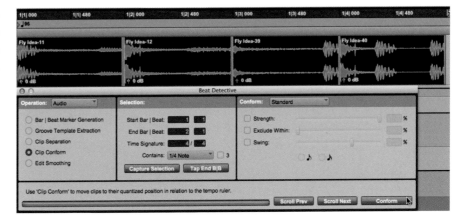

The good news is that Beat Detective offers a 'quick fix' for this! Once you have separated and conformed the clips, you can switch to the Edit Smoothing operation, choose whether to simply fill the gaps (automatically trimming the clip endpoints to fill the gaps) or to fill and crossfade (which also adds a few milliseconds of pre-fade before each clip start point to further reduce the possibility of hearing these edit points). When you click Smooth, everything is adjusted for you automatically – see Figure 3.64 – which saves a lot of editing time if there are lots of gaps to fill.

Figure 3.64
Edit Smoothing using
Beat Detective.

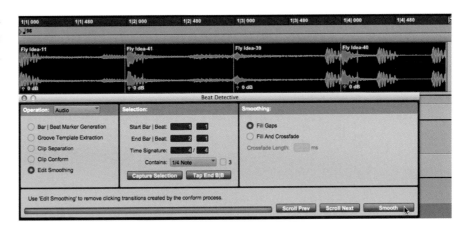

Summary

Having control of the tempo, meter, and key signature of your music is essential if you are to act as the 'conductor' of the musicians, band or orchestra that you are recording – whether real or 'virtual'!

If your role is to engineer the session, then you need to be able to act on the instructions of the producer, arranger, composer, musical director, or of the musicians who are performing on the recording to establish the tempo, meter, and key. This, ideally, should be done before you start recording, or as soon as possible after recording while the people who performed (including any conductor, arranger, or producer directing the performances of the musicians) are still present at the session.

There will be times when the people involved do not want to be constrained by having to play to a click. Everyone needs to clearly understand that any further recording, editing, or mixing of the track will be much more difficult until the correct tempo(s), time signature(s), and key signature(s) have been worked out for the session. If this has to be done after the recording, it can take a significant amount of time to sort all this out. Fortunately, Pro Tools provides excellent tools to help you do this, including Identify Beat, Beat Detective, and Elastic Audio.

You need to be clear about the differences between the sample-based (absolute) and the tick-based (relative) timebases and how these are displayed using the rulers. Also you need to know how and why you can change tracks to be sample based or tick based; the differences between Tempo events and Bar|Beat Markers; when to choose Bar|Beat or Absolute Reference Markers; and when to view the Timeline using either the Linear Sample Display or the Linear Tick Display Mode. You should also be aware of how to carry out Time Operations such as changing the meter, inserting or cutting time from your session, and moving the Song Start.

Understanding the differences between chromatic and diatonic key transposition requires some knowledge of music, but should not be beyond the capabilities of anyone who can master the other intricacies of using the Pro Tools software. Appreciating the differences and what is possible when transposing MIDI and audio clips also requires understanding of music and of MIDI itself. There are many good books published elsewhere that cover these topics more thoroughly than is possible here.

Elastic Audio and Beat Detective are great 'tools' to have at your disposal when you need to make your audio follow the session tempo, or to edit or move notes to align with others, and are well worth spending time with so that you are familiar with all their applications.

Anyway, if you are still unsure about any of these things, please go back and review this chapter, and make sure that you try everything for yourself on your own system. For some people, myself included, it can take a little while before the importance of all these things becomes apparent. When this does happen – learning follows much more quickly!

In this chapter

CHAPTER 4

Working with MIDI

Introduction

Many Pro Tools users will use MIDI in their music production work – even when starting on a project that may end up as audio-only by the time that it is finished. It's just great to be able to try ideas out using piano, bass, drums, percussion, orchestral sounds, or vocal samples when you are working on a new composition or arrangement. Of course, there is plenty of music being recorded today using MIDI instruments, virtual instruments, synthesized and sampled sounds, very heavily, or even exclusively.

Computer MIDI Setup

Before you can use any external MIDI hardware with Pro Tools (such as controllers, keyboards, synthesizers, drum machines, samplers, sequencers, or sound modules), you will need to set up the MIDI connections to and from your computer first.

On Windows computers, you need to use MIDI Studio Setup utility software, and on the Mac, you use the Audio MIDI Setup utility.

On the Mac, you can open the MIDI Studio Setup application using the 'MIDI Studio…' menu option available from the Setup menu in Pro Tools. Here, you can add a device corresponding to each actual device in your MIDI rack and hook these up to your MIDI interface to reflect the actual cable connections you have made.

In this example, I am using a basic setup with just three MIDI devices – an Oxygen 8 keyboard, a Yamaha KX88 keyboard, and a Roland D110 MIDI module – connected to the MIDI interface on a Focusrite Saffire audio/MIDI interface – see Figure 4.1.

On the PC, when you choose MIDI Studio Setup from the Setup menu, you can create an entry for each 'Instrument' that you have in your setup in a similar way. Pop-up selectors let you choose from a list of popular MIDI devices and models, and choose the Input and Output ports. You can also choose which MIDI channels to send and receive on for each device.

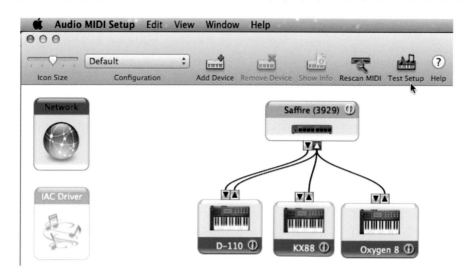

Figure 4.1
MIDI Studio 'page' in the Apple Audio MIDI Setup utility.

TIP

Click on 'Test Setup' in the Apple MIDI Setup MIDI Studio window to activate this feature. To check that the MIDI interface is receiving incoming MIDI data from your MIDI keyboard or controller, just press a key or send a MIDI message, and the 'In' arrow on the interface icon in the MIDI Studio window will light up if it receives this.

To check that all your MIDI devices are receiving MIDI data from the computer, click on each device (synthesizer, MIDI module, etc.) connected to your interface in the MIDI Studio page. This sends a bunch of MIDI notes to the device under test. You should see MIDI Input indicators flash on each device that you test, and you should hear audio if the device's audio outputs are connected to a monitoring system (and don't forget that many synthesizers also have headphone outputs that you can audition to make sure these are playing OK).

Pro Tools MIDI Setup

Once you have connected your external MIDI equipment and have set up your MIDI Studio on your computer, it's time to get Pro Tools configured for MIDI. The first thing to do here is to open the MIDI Input Enable dialog window by choosing 'Input Devices…' from the Setup menu – see Figure 4.2 Here you need to make sure that each device you wish to use to send MIDI to Pro Tools has a tick in the box next to it.

Figure 4.2
The MIDI Input
Enable dialog
window.

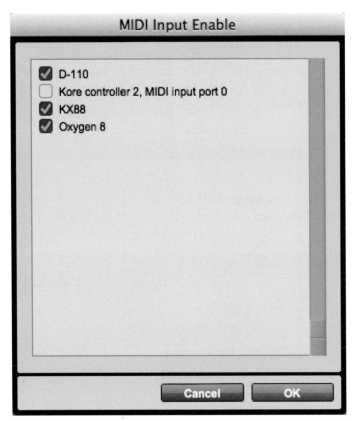

Figure 4.2
The MIDI Input Enable dialog window.

NOTE

All devices correctly configured in your MIDI Studio Setup and physically connected via your hardware interface will be able to receive MIDI data, but MIDI data sent by these devices can only be recorded into Pro Tools if they are enabled in the MIDI Input Enable dialog.

If any of your external (or some virtual/internal) MIDI devices need to receive MIDI Beat Clock messages, this is probably a good time to enable these using the MIDI Beat Clock dialog that is also available from the Setup menu.

You may also need to filter out some of the MIDI data that is being transmitted over your MIDI network using the MIDI Input Filter – also available from the Setup menu. By default, Aftertouch messages are filtered out, but you may decide to exclude other types of messages, such as program changes, depending on the way your MIDI equipment is configured and what you are aiming to achieve.

If you wish, you can set up your Pro Tools session so that incoming MIDI notes are quantized to correct the timing – drum-machine style – using the Input Quantize feature in the Event menu's Event Operations sub-menu – see Figure 4.3. This works well if you are programming drum beats and/or simple bass or synthesizer riffs, and you can preserve as much or as little of the subtleties in the playing depending on the settings you choose here.

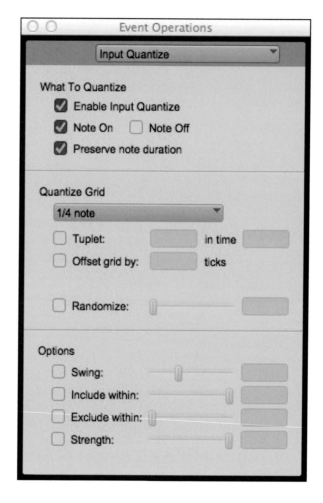

Figure 4.3

The Event Operations Input Quantize dialog.

MIDI Tracks

The most usual way to record MIDI data is to connect an external MIDI keyboard, turn 'Local Control' off if this is a synthesizer or sampler with its own internal sounds, and use this external keyboard to play whichever MIDI device you wish to record.

For example, if you want to play an *external* MIDI device using this keyboard, you need to make sure that MIDI Thru is selected (ticked) in the Options menu, so that the MIDI data transmitted from the keyboard to the computer will be sent back out to the external device. Don't forget: unless the MIDI messages from the keyboard are routed through your computer via the MIDI interface and back out to the external devices, you won't be able to play or hear these!

Before you start recording, you will typically need to set the tempo and meter and set up a click track (see Chapter 3 for more about doing these things). You will probably find it convenient to arrange the Transport window to show the Counters, with these set up to display bars and beats and time elapsed, and, of course, the MIDI controls – as in Figure. 4.4.

Figure 4.4
The transport set up to show the Counters with the Main Counter set to bars and beats and the secondary counter showing elapsed time in hours, minutes, and seconds. The MIDI controls are revealed at the right, and the Record-ready button is enabled.

NOTE
In my simple setup, I am using an Oxygen 8 and a KX88 keyboard. Neither of these has its own sounds, so there is no Local Control to turn off. But to play sounds on my Roland D110 from either of these keyboards, I do need to make sure that MIDI Thru is enabled.

When you have set up your Pro Tools session ready to work with MIDI, go ahead and add one or more MIDI tracks, choosing an available MIDI keyboard as the input and any available (external or virtual/internal) MIDI device as the output. The MIDI Input and Output selectors are located in the track display (Edit window) or channel strip (Mix window) where the Audio Input and Output selectors would be on an Audio track.

Auxiliary Inputs and Audio Tracks for MIDI

You will need to add one or more Auxiliary Input tracks to your session to monitor any external sound sources. You should also add a corresponding number of Audio tracks if you want to record the audio from these sound sources to disk.

To listen to an external MIDI device playing through your Pro Tools mixer, add an Auxiliary Input track and route the audio output from your external MIDI device into this via your audio interface – as in Figure 4.5.

When you are ready to record your MIDI track as audio, use a bus pair to route the output of the Aux track that you are using to monitor your MIDI instrument into the Audio track that you have created to record it – as in Figure 4.6.

Figure 4.5
A Pro Tools Session showing a MIDI track routed to a D110 MIDI module, with an Auxiliary Input to monitor the stereo audio output from the D110 and an Audio track being used to record this audio onto disk.

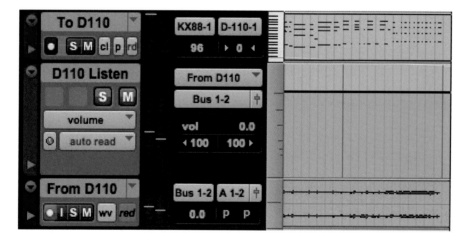

Figure 4.6
Edit window, with Input/Output selectors visible, showing a MIDI track playing notes on a D110 MIDI module; an Auxiliary Track 'listening' to the D110; and an Audio track recording the audio from the D110.

TIP

When you finish recording the MIDI track as audio, you can make the Auxiliary track inactive, mute the MIDI track, hide both of these tracks, and just play back the audio track. Of course, if you need to make any changes at any point in the future, you will still have these original MIDI and Aux tracks available to return to.

Instrument Tracks

Instrument tracks make your session a little more convenient to set up if you are planning to record just one MIDI part for each MIDI or virtual instrument. An Instrument track is a substitute for a MIDI track paired with an Auxiliary Input track used to monitor the audio output from a MIDI hardware instrument or a virtual software instrument. So an Instrument track can do everything a MIDI track needs to do and everything an Auxiliary track needs to do when recording and playing back MIDI data and monitoring the audio data from a MIDI or virtual instrument.

Like a MIDI track, the Instrument track lets you record MIDI data onto the track. Like an Auxiliary Input track, the Instrument track lets you monitor audio coming from an external MIDI device or lets you monitor audio from an inserted virtual instrument plug-in.

If you are recording MIDI drum parts, you may want to record each drum onto a separate MIDI track so that you can edit these separately. So you might have tracks for bass drum, snare drum, hi-hats, toms, and cymbals. If this is the case, there is not much advantage to using an Instrument track: you might as well just use an Auxiliary Input track to monitor the audio from the drum-machine, sampler, or virtual instrument. Of course, you could use an Instrument track for one of these instruments, such as the bass drum, then use separate MIDI tracks for the snare and other drums, with the MIDI output of each track set to control the appropriate drum sound from the MIDI or virtual instrument that is being monitored via the Instrument track.

Recording MIDI

Recording MIDI onto a MIDI or Instrument Track

To record MIDI data, first you need to create a MIDI or Instrument track or choose an existing one. Then assign the track input and output to accept input from the MIDI keyboard or controller that you will be using and route the track's MIDI output to the MIDI device that you wish to control.

In the Mix window, you can click the track's MIDI Input Device/Channel Selector and assign a device and channel from the pop-up menu – see Figure 4.7. If you have more than one keyboard or other MIDI device hooked up that you want to play and create MIDI data with, you can use the 'All' option to allow input from all your connected MIDI devices. I have two MIDI keyboards, for example – one with a weighted action for playing piano sounds and other with a non-weighted action for playing synthesizer sounds.

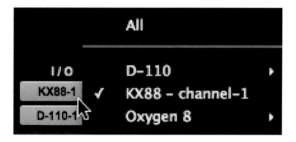

Figure 4.7
The MIDI input pop-up selector showing the available sources in my test session.

You should also assign the MIDI output of the track to the MIDI instrument (the actual hardware or the 'virtual' software) that you will be using, as in Figure 4.8. You may also need to enable the MIDI Thru item in the Options menu.

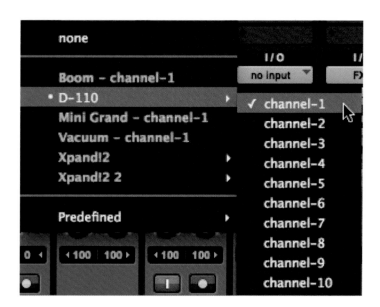

Figure 4.8
The MIDI output pop-up selector showing the available destinations in my test session.

> **TIP**
>
> Remember that when you are recording MIDI onto Instrument tracks, the MIDI Input and Output pop-up selectors are located at the top of the Instruments section rather than in the track I/O section. This is because Instrument tracks can be used to accept audio input from external MIDI devices as well as from virtual instrument plug-ins. So the track I/O section is used to set up the input and output paths for this audio – not to set up MIDI I/O.

Record Mode

To enter Record mode, first you need to Record-enable the track and make sure that it is receiving MIDI from your keyboard or other controller. The track meters will indicate MIDI input when you play your MIDI device if everything is set up and working OK.

In the Transport window, click Return To Zero to start recording from the beginning of the session – or simply press the Return key on the computer's keyboard. If you prefer, you can start recording from wherever the cursor is located in the Edit window. Or, if you select a range of time in the Edit window, recording will start at the beginning of this selection and will automatically finish at the end of the selection.

Click Record in the Transport window to enable Record mode then click Play in the Transport window or press the Spacebar to actually begin recording, as in Figure 4.9.

> **TIP**
>
> A quicker way to start recording is by pressing and holding the Command key (Control key on the PC) then pressing the Spacebar on your computer keyboard. And if you prefer to hit a single key, just press function key F12, or press 3 on the numeric keypad (when the Numeric Keypad Mode is set to Transport) instead.

Figure 4.9
Transport window
with record and play
buttons engaged.

If you enable the Wait for Note button in the Transport window (this is the first of the four buttons in the MIDI controls section and has a MIDI socket icon), the first MIDI note that you play will start the recording. If you want to hear a metronome click while you record, make sure that the click is set up correctly and enable the Metronome button. To hear a countoff, enable the Countoff button. These buttons will each turn blue when enabled.

When you have finished recording, click Stop in the Transport window or press the Spacebar. The newly recorded MIDI data will appear as a MIDI clip on the track in the Edit window and in the Clips List. To see the notes, choose the Notes view (as shown in Figure 4.10) from the pop-up selector in the Track controls area at the left of the Edit window.

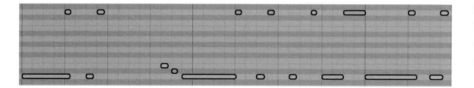

Figure 4.10
MIDI data in notes view in the Edit window.

Cancelling or Undoing Your MIDI Recording

If you mess up while you are recording, which can easily happen before you get used to playing the part you want to record, you can stop the recording and simultaneously discard any stuff that you have recorded (including multiple takes when in Loop Record) using one simple keyboard command. On the Mac, just press the Command key (Mac) or Control key (Windows) together with the period (.) key.

If you have stopped recording and then decide that you want to discard the last take, you can use the standard Undo (last action) command by selecting this from the Edit menu or pressing Command-z (Mac) or Control-z (Windows).

Loop Recording MIDI

There are two ways to Loop Record with MIDI – either using the normal non-destructive Record mode with Loop Playback and MIDI Merge enabled for drum-machine style recording or using the special Loop Record mode to record multiple takes on each record pass – as when loop recording audio.

Drum-machine Style Recording

To set up drum-machine style loop recording, where each time around the loop you record extra beats until you have constructed the pattern you want, you need to enable the MIDI Merge function by clicking its button in the Transport window (see Figure 4.11) or in the Edit window's MIDI Controls section.

Figure 4.11
The MIDI Merge button.

Make sure that 'QuickPunch', 'TrackPunch', 'DestructivePunch', 'Loop Record', and 'Destructive Record' are not selected in the Operations menu, but do select 'Loop Playback', so that the loop symbol appears around the 'Play' button in the Transport window. You should also disable 'Wait for Note' and 'Countoff' in the Transport window. Select 'Link Edit and Timeline Selection' from the Options menu, then make a selection in the Edit window to encompass the range that you want to loop around, as in Figure 4.12.

If you want to hear the audio that plays immediately before the loop range as a cue, you will need to set a pre-roll time. So, for example, if you set a 2-bar pre-roll, you would hear your session play back from 2 bars before the loop range, then it would play around the loop until you hit 'Stop'. Each time through the loop you can add more notes until it sounds the way you want it to, without erasing any of the notes from previous passes through the loop – just like using a drum-machine.

> **TIP**
> To switch to a new record track, press Command (Mac) or Control (Windows), and press the Up/Down Arrow keys to record-enable the previous or next MIDI or Instrument track.

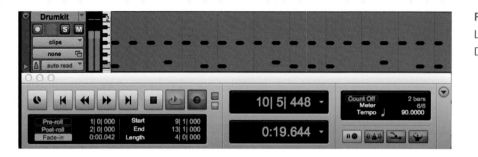

Figure 4.12
Loop Recording –
Drum-machine style.

Loop Record Mode for MIDI

You can record MIDI using the Loop Record mode instead of using the drum-machine style method. In this case, new clips are created each time you record new notes during successive passes through the loop.

To set this up, select 'Loop Record' and 'Link Edit and Timeline Selection' from the Options menu, and de-select 'Loop Playback'. With 'Loop Record' enabled, a loop symbol appears around the 'Record' button in the Transport window. Make a selection in the Edit window to encompass the range that you want to loop around and set a pre-roll time if you need this. Start recording and play your MIDI keyboard or other MIDI controller. Each time around the loop, a new MIDI clip is recorded and placed into the Edit window, replacing the previous clip. When you stop recording, the most recently recorded of these 'takes' is left in the track and all the takes appear as consecutively numbered clips in the MIDI clips list.

The easiest way to audition the various takes is to make sure that the take currently residing in the track is selected, then Command-click (Control-click in Windows) on the selected clip with the Selector tool enabled. A pop-up menu, the 'Takes List', appears containing all your recorded takes – as in Figure 4.13. You can choose any of these to replace the take that currently appears in the track.

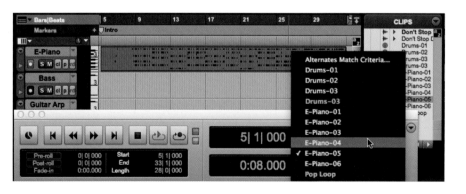

Figure 4.13
Auditioning 'takes' in
loop record mode.
These are also
accessible from
the clips list.

Choosing the Main Ruler

By default, for music sessions the Main Ruler will be set to Bars|Beats. I usually find it helpful to display a secondary ruler showing Min:Secs, or Timecode if I am working to picture. The Main Ruler can be chosen in the area to the left of the rulers in the Edit window by clicking on the downwards-pointing arrow – see Figure 4.16. Whichever ruler is highlighted here becomes the Main Ruler and its data will be displayed in the Main Counter.

Figure 4.16
Choosing the Rulers.

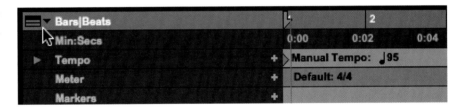

How to Remove Duplicate Notes

One of the most basic, yet useful, MIDI Editing commands is 'Remove Duplicate Notes'. It is very easy to accidentally hit a note on a MIDI keyboard twice – you might hit it tentatively just before it should be played and then again more positively on the beat, or a little late. If a note starts within the first 25% of the duration of a note of the same pitch that is already sounding (or within an eighth note, whichever is shorter), it is considered a duplicate and is combined with the previous note. If it starts later than that, the first note is shortened, so that it ends at the same tick at which the new one starts. To use this command, make an Edit selection that includes the duplicate notes, then choose 'Remove Duplicate Notes' from the Event menu.

How to Deal with Stuck Notes

A typical problem that comes up time and time again when you are programming MIDI instruments is 'stuck notes'. Typically, this happens when you change 'patch' before the patch you are playing has received a MIDI Note Off command. If this happens, you can use the 'All MIDI Notes Off' command from the Event menu or press Command-Shift-Period (.) if you are using a Mac or Control-Shift-Period (.) if you are using Windows. This will send out a series of messages to all the MIDI devices attached to your interface instructing them to turn off any stuck notes.

Mirrored MIDI Editing

Mirrored MIDI Editing lets you edit MIDI clips and have your edits apply to every MIDI clip of the same name. This can be particularly useful when editing looped MIDI clips.

To enable Mirrored MIDI Editing, you can select the 'Mirror MIDI Editing' item from the Options menu. To disable this feature, de-select this menu item. You can also use the Mirrored MIDI Editing button in the Toolbar section of the Edit window to 'toggle' this feature on or off.

To warn you that your edit is being applied to more than one clip, when you make an edit with Mirrored MIDI Editing enabled, the Mirrored MIDI Editing button will blink Red, just once – so keep an eye out for this.

> **TIP**
> It is a good idea to disable Mirrored MIDI Editing when you are not using it, as it is all too easy to inadvertently edit a clip not realizing that this is editing other clips of the same name when you don't want to do this.

Selecting and Playing Notes with the Mini-keyboard

You can use the mini-keyboard at the left of each MIDI or Instrument track in the Edit window to select and play notes by clicking on its keys with the mouse – with any Edit tool selected. When you click a key on the mini-keyboard, all the notes at that pitch on the track immediately become selected, ready for editing.

If you don't want the notes to be selected – because you just want to hear a note at that pitch play – hold the Command and Option keys (Mac) or the Control and Alt keys (Windows) while you click on the mini-keyboard.

If you want to select and play a range of notes, just click a key and drag up or down on the mini-keyboard, and if you want to lengthen or shorten the range of notes selected, you can Shift-click on the mini-keyboard.

> **TIP**
> If you want to de-select notes or select notes that are not adjacent to one another, hold the Command key (Mac) or Control key (Windows) and click on the notes.

'Do to All'

Two extremely useful keyboard commands let you apply changes to all tracks, or to whichever tracks you have selected. For example, maybe you want to set all the track outputs to the same destination ('Do to All'). Or maybe you want to set all *selected* track outputs to the same destination ('Do to All Selected').

To 'Do to All', press and hold Option-click (Mac) or Alt-click (Windows) as you make your change. To 'Do to All Selected', press and hold Option-Shift-click (Mac) or Alt-Shift-click (Windows).

Special Cut, Copy, Paste, and Clear Commands

The Edit menu commands such as Cut Special, Copy Special, Paste Special, and Clear Special can be very useful for editing MIDI controller data. These each have three sub-menu selections two of which let you edit either all the MIDI controller data or just the MIDI pan data. (The third sub-menu command only works with audio plug-in automation.) Conveniently, these commands work whether the data is showing in the track.

The Cut Special, Copy Special, and Clear Special commands work identically. So the 'All Automation' sub-menu selection cuts, copies, or clears all MIDI controller data and the 'Pan Automation' sub-menu selection cuts, copies, or clears only MIDI pan data.

The Paste Special sub-menu commands require further explanation: The 'Merge' sub-menu selection pastes MIDI controller data from the clipboard to the selection and merges it with any current MIDI controller data in the selection.

The 'To Current Automation Type' sub-menu selection pastes MIDI controller data from the clipboard to the selection and changes this to the current MIDI controller data type. This lets you convert from any MIDI controller data type to any other MIDI controller data type. For example, you could copy MIDI volume data and paste this to MIDI pan.

The 'Repeat to Fill Selection' sub-menu selection repeatedly pastes the MIDI controller data from the clipboard until it fills the selection – saving lots of time compared with manual pasting of selections. Simply cut or copy a MIDI clip, then make an Edit selection and use this command to fill the selection. If you have selected an area that is not an exact multiple of the copied clip size, it will automatically trim the last copied clip that is pasted so that it fits exactly.

Graphic Editing in the Edit Window

Pro Tools MIDI tracks can be edited graphically in the Edit window. Here you can use the standard Pro Tools Trimmer tool to make notes shorter or longer and use the Grabber tool to move the pitch or position – or 'draw' notes in using the Pencil tool. You can draw in or edit existing velocity, volume, pan, mute, pitch bend, aftertouch, and any continuous controller data, and the Pencil tool can be set to draw freehand or to automatically draw straight lines, triangles, squares, or randomly. The Pencil tool also lets you draw and trim MIDI note and controller data, and the Trim tool can trim MIDI note durations when a MIDI track is set to Velocity view.

It can be very handy at times to insert notes using the Pencil tool instead of setting up an external keyboard. Just make sure that the MIDI track is in Notes view and select the Pencil tool at the top of the Edit window. To insert quarter notes on the beat, for example, set the Time Scale to Bars and Beats, then set the Edit mode to Grid and the Grid value to quarter notes. As you move the Pencil tool vertically and horizontally within a MIDI track, the bar/beat/clock location and the pitch of the MIDI note are shown in the Cursor Location and Cursor Value displays just below the Main Counter in the Edit window toolbar. When you find the note and position you want, you can just click in the track to insert it. It's as simple as that!

To select notes for editing, you can either use the Grabber tool to drag a marquee around the notes, or drag using the Selector tool across a range of notes. Once some notes are selected, you can drag these up or down to change the pitch using the Grabber or Pencil tools – while pressing the Shift key if you want to make sure that you don't inadvertently move the position of the notes in the bar. You can use the Trimmer tool or the Pencil tool to adjust the start and end points of the notes. If you set the MIDI track to Velocity view, you will see the attack velocities of the notes represented by 'stalks'. You can edit these using the Grabber tool.

Sometimes, you may simply need to edit one note. If you select a note using the Grabber or the Pencil, its attributes will be displayed in the Event Edit area to the right of the counters in the toolbar at the top of the Edit window.

Here you can type in new values for any of the displayed parameters. See the accompanying screenshot, Figure 4.17, for more details.

Figure 4.17
The selected MIDI note's attributes are displayed in the Event Edit area to the right of the Counters, ready for editing. The Main Counter shows the bar/beat/clock location of the start point of the selected MIDI note. The bar:beat:tick location of the cursor (109:3:316) and the pitch of the MIDI note that the cursor is pointing to (A1) are shown in the Cursor Location and Cursor Value displays just below the Main Counter.

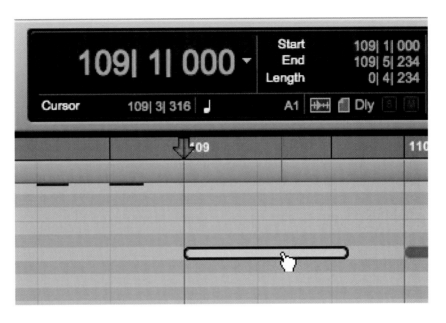

Track Views

You can switch the Track View of any MIDI track in the Edit window to show MIDI Continuous Controller data. When continuous controller data is recorded onto MIDI tracks, it is displayed as a line graph with a series of editable breakpoints. These breakpoints are stepped to represent individual controller events, as in Figure 4.18, in contrast to the standard automation breakpoints, which are interruptions on a continuous line.

Figure 4.18
MIDI Continuous Controller data line graph with stepped breakpoints.

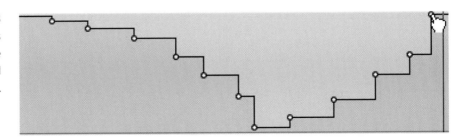

The Track View for Instrument tracks can either show MIDI CC data or show audio automation data, such as audio volume – see Figure 4.19.

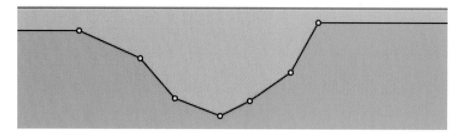

Figure 4.19
Audio Automation data vector graph with breakpoints on continuous line.

You can edit MIDI Continuous Controller data directly in the Edit window according to which type of data you select using the pop-up Track View selector – see Figure 4.20. As well as blocks, clips, and notes views, the Track View selector for MIDI tracks lets you choose velocity, MIDI volume, mute, pan, pitch bend, mono aftertouch, program change, or SysEx, and lets you access other controller views using the controllers sub-menu.

The Track View selector for Instrument Tracks (see Figure 4.21) additionally lets you display audio volume and volume trim, audio mute, pan left and right, and any send or plug-in automation data.

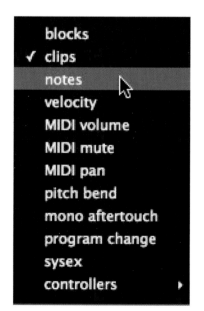

Figure 4.20
MIDI Track View selector.

Figure 4.21
Instrument Track View
selector.

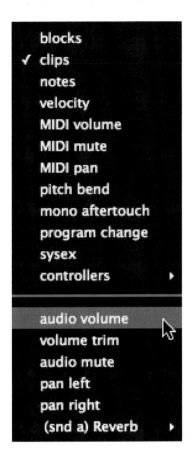

For example, with MIDI volume selected in the Track View, you can use the Pencil tool to draw MIDI volume data – as in Figure 4.22.

Automation Lanes

You can display the automation data more clearly in a separate lane below the notes display in the Edit window. Click on the disclosing arrow at the bottom left of the Track controls (see Figure 4.23) to show the automation lanes and choose the type of automation that you want to display using the pop-up selector provided. You can add more automation lanes using the small '+' icon, and you can remove these using the small '–' icon in the controls area at the left of each automation lane.

In Velocity view, you can edit each note's attack velocity using the Grabber tool by dragging its velocity stalk upwards or downward: the taller the stalk, the higher the velocity value – see Figure 4.24. You can use velocities to create crescendos or to make accents within MIDI parts.

Figure 4.22
Drawing MIDI volume data in the Edit window on a MIDI track using the Pencil tool.

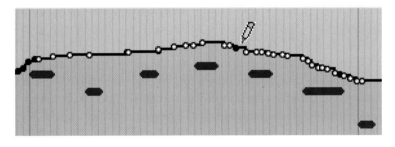

Figure 4.23
Revealing the automation data in a separate lane below the note data can make this easier to edit.

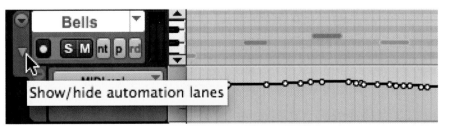

183

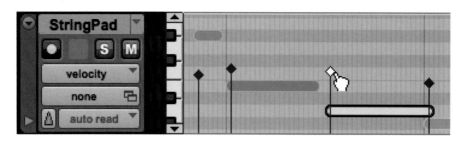

Figure 4.24
Velocity data shown in an Instrument track overlaid on top of the note data.

It can sometimes easier to edit the velocity data if you open a separate lane for this – see Figure 4.25.

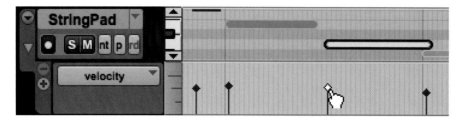

Figure 4.25
A lane opened below the MIDI/Instrument track to allow Velocity editing.

> **NOTE**
> MIDI controller #7 (volume) and #10 (pan) are treated by Pro Tools as automation data, so these controller events (along with Mutes) can be recorded and automated from the Mix window. Consequently, a MIDI or Instrument track's automation mode will affect how these events are played back and recorded. For example, if you suspend automation for a MIDI track, the MIDI volume, pan, and mute events will be suspended, but any other MIDI controller events will continue to operate. To avoid confusion, remember that Instrument tracks also support audio volume, pan, and mute along with MIDI volume, pan, and mute.

Note and Controller Chasing

New MIDI and Instrument tracks have Note Chasing enabled by default, but this can turned on or off individually for each MIDI or Instrument track from the pop-up Playlist selector menu – see Figure 4.26.

Figure 4.26
Enabling Note
Chasing using the
Playlist selector
pop-up menu.

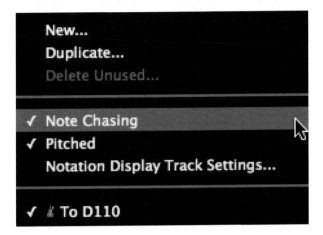

Note Chasing is necessary to allow sustained MIDI notes to be heard when you start playback at some point after the note's start time. If a long note started playing back at the start of the chorus at, say, bar 9 and lasted for 8 bars, you would not hear this note if you started playback half-way through this chorus section at bar 13 – unless, of course, Note Chasing is enabled.

> **NOTE**
> You should disable Note Chasing when you are working with samplers that are playing loops because Note Chasing would trigger the loop to start playback from wherever you start the session playback – inevitably putting the loop out of time with the other tracks unless you start exactly at the correct loop trigger location.

As far as continuous controller events and program changes are concerned, Pro Tools always chases these to make sure that controller values and any patches for MIDI devices are always set correctly.

Event List Editing

You can also view and edit MIDI data using the MIDI Event List, which is available from the Event menu. This lists MIDI events alphanumerically with letter names for the notes and numbers for the locations, the On and Off velocities, and the note lengths. In this window, you can select, copy, paste, or delete events and you can edit values by typing directly in the list. A pop-up selector at the top left of the window lets you select which MIDI or Instrument track to display in the event list – see Figure 4.27.

Figure 4.27
Data in the MIDI Event List window.

A second pop-up at the top right of the window has various commands that let you customize the MIDI Event List, make it scroll the way you want it to, or choose where events are to be inserted. You can also choose what to display in the list using the View Filter. A sub-menu, labelled Insert, lets you insert any MIDI event (except SysEx) into the list – see Figure 4.28.

> **TIP**
> When the MIDI Event List is the front-most window, you can simply press Command-N (Mac) or Control-N (Windows) to Insert a note, and use the left/right arrows on your computer keyboard to move between the Event Entry fields. To insert a Controller Event, press Command-L (Mac) or Control-L (Windows), and to insert a Program Change Event, press Command-P (Mac) or Control-P (Windows). After inserting any of these events, you can insert another similar event by pressing Command-M (Mac) or Control-M (Windows).

Figure 4.28
MIDI Event List
Options pop-up
selector showing
Insert sub-menu.

NOTE

Pro Tools has no list editor for audio event data – a major omission in my opinion.

Quantize to Grid Command

The Quantize to Grid command (see Figure 4.29) adjusts the placement of selected audio and MIDI clips (but not individual MIDI notes or individual Elastic Audio Events), so that their start points (or sync points, if present) align precisely to the nearest Grid boundary.

The way this works is that you choose a Grid value using the controls in the Edit window toolbar, use the Selector or Time Grabber to select the clip or clips (on one or more tracks) that you want to quantize, then choose the 'Quantize

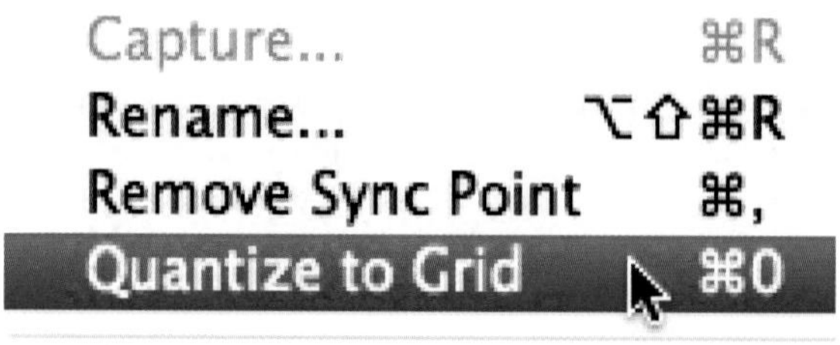

Capture...	⌘R
Rename...	⌥⇧⌘R
Remove Sync Point	⌘,
Quantize to Grid	⌘0

Figure 4.29
Part of the Clip menu showing the Quantize to Grid command.

to Grid' command from the Clip menu – or press Command (Mac) or Control (Windows) and '0' on the numeric keypad. The Clip start times (or sync points if these are different) will then be automatically aligned with the nearest Grid lines.

With MIDI clips, only the clip start times or sync points are quantized: all the MIDI data within the clips (such as notes) is moved equally, thereby retaining its rhythmic relationships. Similarly, with Elastic Audio clips, only the clip start times or sync points are quantized: all Elastic Audio Events within the clips are moved equally, thereby retaining their rhythmic relationships.

> **NOTE**
> The Event Operation Quantize features can be applied to MIDI notes and Elastic Audio Events individually in addition to audio clip start times (or sync points).

MIDI and Instrument Controls and Track Heights

Each MIDI or Instrument track has a set of controls that includes the Track Name, buttons for Record-Enable, Solo, Mute, and Patch Select, a Track View selector, a Playlist selector, a Track Timebase selector, an Automation Mode selector, and a Track Height selector.

To change the track heights for MIDI or Instrument tracks, you can click on the mini-keyboard at the left of the Track display in the Edit window and select the height you prefer – see Figure 4.30.

NOTE

When the MIDI or Instrument track is in Notes or Velocity view, you will need to Control-click on the mini-keyboard to access the track heights pop-up.

Figure 4.30
MIDI Track Height
pop-up showing the
available options
(micro to extreme),
including the single
note options.

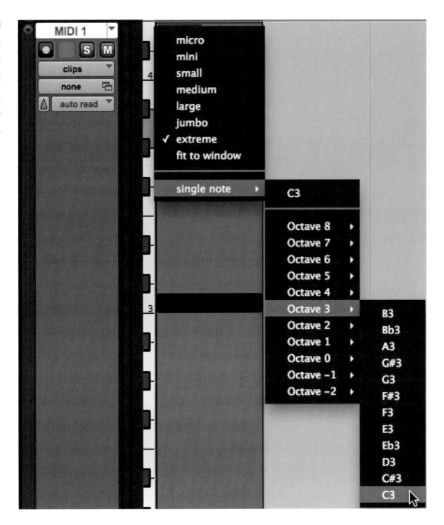

It can be useful to view just one particular note pitch within a MIDI or Instrument track, such as those that play a single MIDI note on a sampler, or when you have split the different instruments of a drum kit onto individual MIDI tracks. Pro Tools provides a Single Note height to use in these situations – see Figure 4.31.

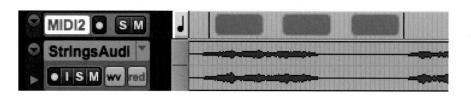

Figure 4.31
A MIDI track set to
Single Note Height.

You can also access the Track Height selector from the Track Options pop-up – see Figures 4.32 and 4.33. When the Track Height is set to small, mini, or micro, the Track Options pop-up lets you access the Track Timebase and Track View options as well – because the track controls area becomes too small to provide space for these other pop-up selectors in their normal positions.

Figure 4.32
Track Options
selector.

Figure 4.33
Track Options pop-up.

In small view, there are three very small buttons provided (see Figure 4.34) to let you access the Track View, Patch Select, and Automation Mode selectors, but even these disappear in mini and micro views.

MIDI Real-Time Properties

The MIDI Real-Time Properties are Quantize, Duration, Velocity, Transpose, and Delay, all of which are attributes of MIDI notes that can be changed by selecting the notes and using the appropriate command from the Event menu.

Figure 4.34
Small Track View
showing three tiny
buttons to the right
of the Mute button
for the Track View,
Patch Select, and
Automation Mode
pop-up selectors.

Delay is slightly different in that this can be achieved in a number of ways, one of which is by entering a MIDI Track Offset from the Event menu. Using the Event menu commands changes the MIDI data, whereas using the MIDI Real-Time Properties feature just alters the data as it is being played back – in real time – leaving the original data unaltered.

Why would you want to use Real-Time Properties instead of using Event menu commands to change the data? This is useful when you want to experiment with different settings before you are sure which you want to commit to. You can even change the Real-Time Properties parameters while the session is playing back, so that you can immediately hear the effect this will have. And when you have made up your mind, there is a command in the Track menu, 'Write MIDI Real-Time Properties' that lets you write these properties to the track – permanently changing the data in the track.

Real-Time Properties for the whole track can either be adjusted in the Real-Time Properties View in the Edit window – or by using the Real-Time Properties window that you can access from the Event menu.

You can also apply MIDI Real-Time Properties to individual clips by selecting the clip or clips and opening the MIDI Real-Time Properties dialog from the Event menu.

> **NOTE**
> Clip-based Real-Time Properties can only be applied using the Real-Time Properties dialog – not by using the Real-Time Properties View in the Edit window.

You will see the letter 'T' displayed in the upper right-hand corner of the track's MIDI clips and in the Event List whenever Real-Time Properties are being applied to a track – as shown in Figure 4.35.

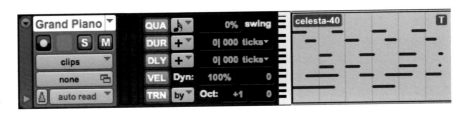

Figure 4.35
Edit window showing the MIDI Real-Time Properties for a track. This track has just one clip, identified with a 'T' in the top right-hand corner to indicate that track-based Real-Time Properties are being applied.

Similarly, you will see the letter 'R' displayed in the upper right-hand corner of the clip (see Figure 4.36) and in the Event List whenever Real-Time Properties are being applied to a clip.

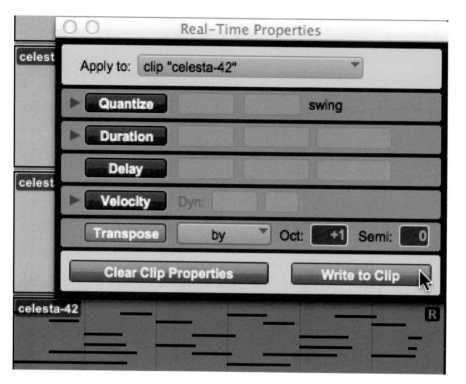

Figure 4.36
The MIDI Real-Time
Properties dialog.
Below this, you can
see a MIDI clip with
the letter 'R' displayed
in the top right-hand
corner to indicate that
clip-based MIDI
Real-Time Properties
are being applied
to this clip.

NOTE

If you wish, you can permanently change the data in the selected clip,
or clips, by clicking on the 'Write To Clip' button in the MIDI Real-Time
Properties dialog.

So, the main reason for using MIDI Real-Time Properties is that you can change
any of the parameters without having to commit to your changes until you are
ready to do this. Also, if you are applying these to a track, you can keep these
properties conveniently displayed in the Edit window if you like, which can be
a big help when you are working out what to do with your MIDI tracks. And
when you are sure that you have the settings you want, you can write these
settings to the selected tracks or clips.

Exporting MIDI Tracks

There will be times when you want to Export MIDI tracks (or Instrument
tracks containing MIDI data) from a Pro Tools session. For example, if you have

recorded SysEx data containing patch information for your MIDI synthesizers or other external MIDI devices into Pro Tools tracks, you may wish to store this data separately for use with Pro Tools or other software in the future. Or you may have developed useful MIDI sequences that you want to be able to build libraries of to use with Pro Tools or other software in the future. If you have recorded an arrangement using MIDI tracks in Pro Tools, you may wish to transfer this to other software, such as Digital Performer, Cubase, Nuendo, or Logic, so that you can work on the MIDI arrangement further using these. Or you may just want to transfer individual tracks to a notation software application such as Sibelius, so that you can print out parts for musicians to play.

For these purposes, Pro Tools lets you export your Pro Tools session, or one or more MIDI or Instrument tracks, as a Type 1 (multi-track) or Type 0 (merged) Standard MIDI file. The MIDI file contains any meter and tempo information and any key signatures used in your session, along with track names and markers.

To export a selected track or a group of selected tracks, choose the 'Export MIDI' command from the File Menu in Pro Tools. The Export MIDI dialog allows you to choose the MIDI file format and the location reference (e.g. Session Start) and has a check box that you can tick if you want to apply Real-Time Properties to the exported data – see Figure 4.37.

Figure 4.37
Export MIDI dialog.

To export a single MIDI or Instrument track as a MIDI file, you can Control-click (Mac) or Right-click (Windows) on the name of the track you want to export and choose 'Export MIDI…' from the pop-up menu that appears – see Figure 4.38. This shortcut also works with a group of selected tracks if you Control-click (Mac) or Right-click (Windows) on the name of any one of them.

Figure 4.38
Track Name pop-up.

The MIDI Editors

Pro Tools provides separate MIDI Editor windows that can be better to use for detailed MIDI Editing than the Edit window. The notes are displayed in a clearer way, for example, with velocity, controller, and automation displayed in separate lanes below the notes display.

To open a MIDI Editor, select any MIDI clip and choose MIDI Editor from the Window menu – see Figure 4.39. You can also open a MIDI Editor for any MIDI clip by double-clicking on the clip when the preference for this is set appropriately in the MIDI Preferences window. You can also use the keyboard command Control-equals (=) on the Mac or Start-equals (=) on Windows.

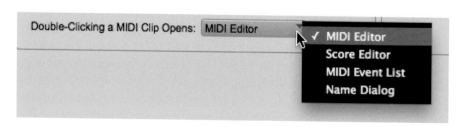

Figure 4.39
Part of the MIDI Preferences window showing what may be opened when double-clicking a MIDI clip.

194

Tracks and Groups in the MIDI Editors

You can view the Tracks and Groups lists at the left of any MIDI Editor window and use this to reveal or hide tracks and to enable or disable Groups, as necessary. To open the Tracks and Groups list, you can either click on the small arrow at the bottom left of the MIDI Editor window (see Figure 4.40), or select 'Track List' from the Toolbar menu at the top right of the MIDI Editor window.

Figure 4.40
The MIDI Editor window showing all the Edit tools and the expanded Grid/Nudge Display in the Toolbar and with the Tracks and Groups Lists open at the left.

By default, MIDI, Instrument, and Auxiliary Input tracks are all shown in the Tracks List. If you prefer, you can show only the MIDI tracks or only the Instrument tracks using the options in the Tracks List pop-up menu – see Figure 4.41.

Figure 4.41
Using the Tracks List pop-up menu to choose what to show in the list.

How the MIDI Editors Work

Each MIDI Editor window has a small red 'Target' button near the top right of the window. If you click this, it will turn grey. Then, if you open another MIDI Editor window, whether this is for the same MIDI clip on the same track or for a different MIDI clip on the same or on another track, this will be opened in a separate window. Otherwise, with the Target button showing red, the new clip will open in place of the existing MIDI Editor window.

If you choose to display multiple tracks in a single MIDI Editor window, the MIDI notes from the different tracks will be superimposed in the window.

Velocity and other continuous controller data for MIDI and Instrument tracks, and automation data for Instrument and Auxiliary Input tracks, is displayed in Controller lanes below the Notes pane in the MIDI Editor windows.

Velocities for MIDI notes are superimposed in the single Velocity lane, while other automation and controller data is shown using individual lanes for each type of automation or controller.

Using the MIDI Editor to Insert Notes

As with the main Edit window and the Score Editor, you can use the Pencil tool to insert notes onto any track in the MIDI Editor. However, because notes for multiple tracks can be displayed simultaneously in a single MIDI Editor window, Pro Tools has a system that lets you choose which of these to edit, preventing you from inadvertently making changes to the others.

Here's how this works: In the Tracks List, there is a column at the right that displays a tiny Pencil icon next to whichever track is currently 'pencil-enabled' to select this for editing – see Figure 4.42.

In the column at the far left of the Tracks List, you can click on the grey dots to show or hide individual tracks within the MIDI Editor window. If the dot is dark grey, the track will be shown; if the dot is light grey, the track will be hidden.

If you only have one track showing in the MIDI Editor, that track will automatically

Figure 4.42
Tracks list showing the tiny Pencil icon that appears to the right of the Track Name when the track is 'pencil-enabled' to allow editing.

Figure 4.43
Pencil-Enabled Track
pop-up selector.

be pencil-enabled. If you have more than one track, you can choose which to enable by clicking to the right of the Track Name in the Tracks List.

The MIDI Editor also has a pop-up selector in its Toolbar area (see Figure 4.43) that provides a quick way to enable a track for editing – particularly handy when the Tracks List is not open. This pop-up lists all the MIDI and Instrument tracks that are visible in the Tracks List, and when you select one of these tracks, this becomes enabled for editing.

If the Tracks List is open at the left of the window, you can simply click in the rightmost column to pencil-enable any of the tracks that are currently showing – and you can show or hide any tracks by clicking in the leftmost column – see the example in Figure 4.44.

Figure 4.44
Pencil-enabling a
track, ready to edit
notes on this track in
the Notes pane.

If you want to be able to edit more than one track, you can Shift-click in the Pencil column to enable a contiguous selection of tracks (i.e. a selection of tracks that are next to each other). If any of the tracks in this range are not already being shown in the MIDI Editor, this action will cause them to be shown as well as pencil-enabled.

If the tracks you want to pencil-enable are not next to each other in the list, you can Command-click (Mac) or Control-click (Windows) in the Pencil column to enable additional tracks. And if you apply this command to tracks that are not currently shown in the Notes pane, they will be both 'shown' and 'pencil-enabled'. If you want to both show and pencil-enable *all* tracks in the Tracks List, the shortcut is to Option-click (Mac) or Alt-click (Windows).

Creating MIDI Notes with the Pencil Tool

To insert new notes, simply choose the Pencil tool from the Toolbar and click in the Notes pane at any time and pitch location. This action will insert a MIDI note on the pencil-enabled track. This will have the Default Note On Velocity unless you change this velocity first by entering a new value for Note On Velocity using the MIDI Editing Controls display. The note entered will also use the Default Note Duration shown in the MIDI Editing Controls, unless you change this first.

Automation Lanes in MIDI Editor Windows

MIDI Editor windows also let you edit velocities, MIDI controller data, and automation for all shown MIDI, Instrument, and Auxiliary Input tracks in lanes under the Notes pane. To show or hide lanes under the Notes pane, just click the Show/Hide Lanes button. This is a small revealing arrow to the left of the keyboard at the bottom left of the Notes pane – see Figure 4.45.

Figure 4.45
Using the Show/Hide Lanes button.

If there are multiple MIDI and Instrument tracks showing in the Notes pane, the velocities for notes on separate tracks will all be shown together in a single lane, superimposed just like notes in the Notes pane – see Figure 4.46.

However, all other Controller lanes are grouped by automation and controller type and provide individual lanes for each shown track. To add or remove lanes, you can use the '+' or '−' buttons at the left of the lane – see Figure 4.47.

Figure 4.46
Superimposed Notes
and Controllers.

Figure 4.47
Using the Add Lane
button.

NOTE
If the vertical zoom for the controller lanes is set to the minimum size, you won't be able to see the Add Lane button. In this case, when you use the pop-up Track Options selector, (see Figure 4.48) the option to Show Automation Lanes will be available, along with sub-menus for Track Height, Playlist, Track Timebase, and Track View.

Figure 4.48
Using the pop-up
Track Options selector.

Each Lane has a pop-up Lane View Selector that you can use to choose which type of data you wish to display in this lane – see Figure 4.49.

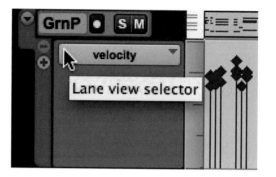

Figure 4.49
Using the Lane View Selector.

You can use the pop-up Lane View selector (see Figure 4.50) whenever you like to choose the Automation or Controller type that you want to display. MIDI tracks let you choose from the standard list of MIDI controllers while Instrument tracks also let you choose from the available audio automation types.

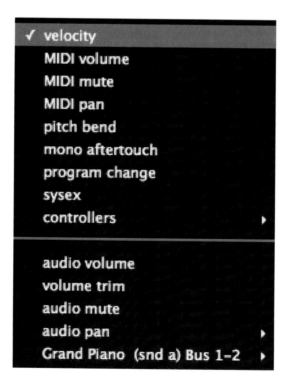

Figure 4.50
Pop-up Lane View Selector showing available MIDI controller and audio automation types.

If you choose MIDI volume, for example, you can click using the Grabber tool to insert breakpoints, dragging these into position or moving existing ones to new positions, or draw in breakpoints using the Pencil tool to create and edit automation data as in Figure 4.51.

Figure 4.51
Editing MIDI volume controller data using the lanes in the MIDI Editor window.

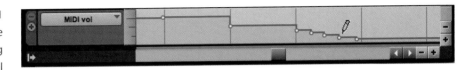

The MIDI Editor Toolbar

Running along the top of the MIDI Editor window you will find a Toolbar which is similar to the Edit window toolbar, but with a number of differences.

Figure 4.52
Solo and Mute buttons engaged and applying to all tracks currently displayed in the MIDI Editor window – normally you would engage one or other of these.

Mute and Solo

At the far left, there are two small buttons for Solo and Mute – see Figure 4.52. The Solo button lets you solo all the tracks currently displayed in the MIDI Editor window, while the Mute button lets you mute all the tracks currently displayed in the MIDI Editor window.

If you have more than one track showing in the MIDI Editor window, engaging the Solo or Mute button in the MIDI Editor solos or mutes all these tracks. If, on the other hand, you solo or mute some, but not all, of the tracks shown in the MIDI Editor window using Solo and Mute buttons in the Edit or Mix windows, the appearance of the Solo and Mute buttons updates to indicate mixed states – see Figures 4.53 and 4.54. Mixed solo and mute states are said to occur when one or more tracks shown in the MIDI Editor window, but not all, are soloed or muted.

Figure 4.53
Mute status indicates a mixed state, revealing that not all the selected tracks are muted.

Figure 4.54
Solo status indicates a mixed state, revealing that not all the selected tracks are solo'ed.

Notation View

To the right of the Solo and Mute buttons, there is a button that lets you switch the note display to show the MIDI notes as music notation on a stave that runs from left to right along the timeline – see Figure 4.55. Each MIDI and Instrument track is displayed independently in Notation view, with one track per staff – they are not superimposed.

201

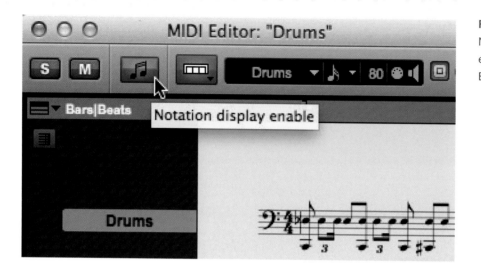

Figure 4.55
Notation display enabled in the MIDI Editor window.

> **NOTE**
> There is also a Score Editor in Pro Tools that displays the notation in a 'page' view instead of as a timeline that runs from left to right.

You can edit notes in Notation view just like you can in the Score Editor (as explained in the next section), but with the advantage of having access to Velocity, Controller, and Automation lanes.

> **NOTE**
> In Notation view, the Trimmer tool functions as the Grabber tool when you hover this above a note and functions as the Note Selector tool when it is not over a note. Also, the Edit mode is automatically set to Grid and cannot be changed.

At the top left of the Notation view, just below the rulers, there is a Double Bar button. If you enable this button, a double bar line will be displayed at the end of the last MIDI clip or event in the session – see Figures 4.56 and 4.57.

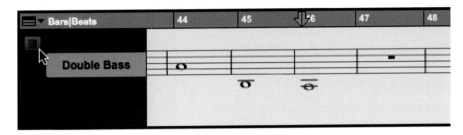

Figure 4.56
Clicking on the
Double Bar button in
Notation view.

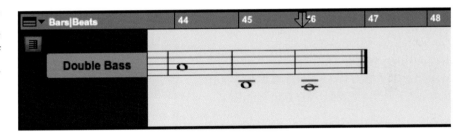

Figure 4.57
A double bar line
indicates the end of
the piece of music.

When this is disabled, there will be a number of empty bars at the end of the last event in the session. You may want additional empty bars to be available, so that you can manually enter new notes after the last MIDI clip or event in the session if you have not yet completed your composition. In this case, you should disable the Double Bar button, and you can specify the number of empty bars using the 'Additional Empty Bars In The Score Editor' setting in the MIDI Preferences page that you can access from the Setup menu.

The Edit Tools

When the Expanded Edit Tools are *not* displayed in the MIDI Editor window, the button to the right of the Notation button enables you to access a popup selector that can be used to define the function of this button. By default, the Zoomer tool is selected – see Figure 4.58.

Figure 4.58
Edit Tool pop-up
selector – defaulted
to the Zoomer Tool.

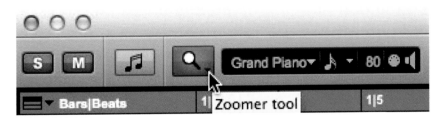

If you click and hold this button, a pop-up selector opens from which you can choose an Edit tool or the Smart tool – see Figure 4.59.

Expanded Edit Tools

You can expand this section of the Toolbar to show these Edit tools as a row of buttons, just like in the Edit window. To open the Expanded Edit tools, click on the pop-up Toolbar menu at the top right of MIDI Editor window – see Figure 4.60.

The Edit tools in the MIDI Editor window (see Figure 4.61) let you work with MIDI data in the same ways as the tools in the Edit window.

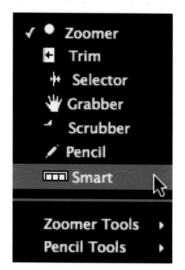

Figure 4.59
Edit tools pop-up selector.

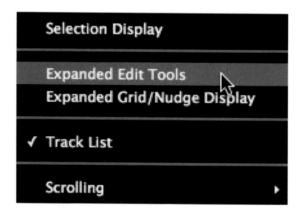

Figure 4.60
Toolbar Menu.

Figure 4.61
The Expanded MIDI Editor tools.

> **NOTE**
> The Edit tool settings can be set differently for each window when you have more than one MIDI Editor window open.

MIDI Editing Controls

To the right of the Edit tools, you will find a set of MIDI Editing Controls that work exactly like the MIDI Editing Controls in the Edit window.

The leftmost of these controls is the Track Edit pop-up Selector that lets you choose which of the visible tracks to edit – see Figure 4.62. Avid calls this 'pencil-enabled' and warns that you can only insert MIDI notes manually in 'pencil-enabled' tracks.

Figure 4.62
The MIDI Editing
Controls.

Next along is the Default Note Duration selector that you will use to choose note durations (see Figure 4.63) when you are manually inserting MIDI notes. A useful choice here is to follow the durations set using the Grid.

Figure 4.63
Default Note Duration
pop-up selector.

The numerical entry field (see Figure 4.64) with the number 95 set by default is the Default Note On Velocity setting for MIDI notes that you manually insert. You can type any number between 1 and 127 here.

The combined MIDI socket and loudspeaker icon is actually a button that, when enabled, causes MIDI notes to play when you insert them using the Pencil tool, click them with the Grabber tool, Tab to them, or select them with the Note Selector tool.

Additional Toolbar Buttons, Controls, and Displays

To the right of the MIDI Editing Controls, there are two more large buttons – see Figure 4.65. The first of these is the Link Timeline and Edit Selection button that should be familiar to you from the Edit window. If you disable this, you can make independent Timeline and Edit selections. The second of these buttons lets you enable or disable Mirrored MIDI Editing. This lets you edit one MIDI clip and have your edits apply to every copy of that same MIDI clip.

Figure 4.65
The 'Mirrored MIDI Editing' and 'Link Timeline and Edit Selection' buttons.

Moving further along the Toolbar you will find the standard Edit Mode buttons, (Shuffle, Spot, Slip and Grid), which operate in exactly the same way as those in the Edit window. There is also a Cursor Location and Cursor Value display that shows the exact location that the cursor is pointing at in the MIDI Editor window and also displays the pitch (value) of the note that the cursor is currently pointing at – see Figure 4.66.

Figure 4.66
The Edit Mode
buttons with the
Cursor Location and
Value Display to the
right of these showing
the position of the
cursor and the pitch
of the note that the
cursor is pointing at.

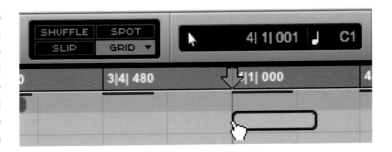

Moving along the Toolbar to the right again, controls for Grid and Nudge can
be displayed either in a compact or expanded format. By default, the compact
form has a popup selector that lets you choose whether to display the Grid or
the Nudge values for editing – see Figure 4.67.

Figure 4.67
The compact version
of the Grid and Nudge
controls, with the
cursor pointing to the
pop-up Grid/Nudge
Toggle selector that lets
you choose which to
control.

Using the Toolbar menu located at the top right of the MIDI Editor window,
you can select 'Expanded Edit Tools', which causes the Grid and Nudge controls
to be displayed separately, side-by-side - as in Figure 4.68.

Figure 4.68
Expanded Grid and
Nudge controls with
the enabled Target
button and the
Toolbar menu pop-up
selector to the right
of these.

To the right of these controls, you can see the Target button, which is coloured red when it is enabled, and the Toolbar menu pop-up selector.

The Target button works the same way that the Target button works in Plug-in, Output, and Send windows, allowing you to have un-targeted windows open at the same time. When you un-target any of these windows by clicking on its red Target button, this button becomes greyed out to confirm that it is un-targeted. The Targeted MIDI Editor window (you can only target one MIDI Editor window at a time) also synchronizes its Timeline location view to the Timeline location view in the Edit window. Consequently, if you make or change an Edit selection in the Edit window, this same Edit selection will be made or changed in the Targeted MIDI Editor window.

The MIDI Editor window's Toolbar menu allows you to open the Track List at the left of the MIDI Editor window by selecting the Track List item and the Scrolling item has a sub-menu that lets you choose from various scrolling options for the MIDI Editor window – see Figure 4.69.

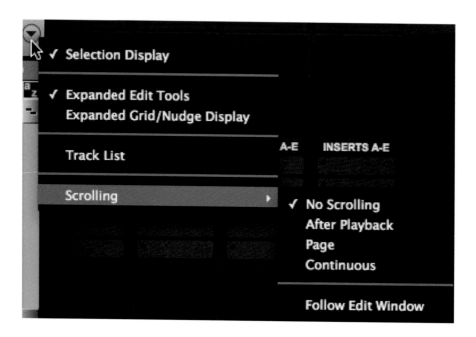

Figure 4.69
MIDI Editor Toolbar menu Scrolling Options.

These Scrolling Options can be set independently for each MIDI Editor window and operate independently from those chosen for the Edit window. There is also an option to have the MIDI Editor window follow the Scrolling settings that you have made for the Edit window.

> **NOTE**
> You can re-arrange the positions of the controls and displays in the toolbar, just as you can in the Edit window: Command-click (Mac) or Control-click (Windows) and simply drag the controls or displays that you want to a new location in the Toolbar.

The MIDI Editor Zoom Controls

There is a Horizontal Zoom control for the MIDI Editor window, consisting of a pair of '+' and '−' buttons, tucked away at the bottom-right corner of the window, just to the right of the scrollbar that runs along the bottom of the window. This controls the vertical zoom for both the Notes pane and the Controller lanes.

There is a Vertical Zoom button for the Notes pane located in the upper-right corner of the MIDI Editor window, just above the right-hand scroll bar. Clicking the top part of this button will increase the vertical zoom, and clicking the lower half will decrease the vertical zoom for MIDI notes.

A second pair of '+' and '−' buttons located just beneath the scrollbar at the right-hand side of the window acts as a Vertical Zoom control for the Automation and Controller lanes that run below the Notes pane.

Timebase and Conductor Rulers

You can choose which Timebase and Conductor rulers to view in each MIDI Editor window independently of each other and independently of the main Edit window, using the pop-up selector just underneath the Solo and Mute buttons – see Figure 4.70. You can also view different time locations from the main Edit window in one or more MIDI Editor windows, and these can be set to different zoom levels.

Figure 4.70
MIDI Editor window showing Timebase and Conductor rulers.

The Superimposed Notes View

The MIDI Editor windows allow you to view and edit MIDI notes from different MIDI and Instrument tracks at the same time, all together, superimposed in the same window. To distinguish notes from different tracks, these can be colour coded by Track or by Velocity using two buttons at the top left of the MIDI Editor window, just below the rulers.

The first button lets you enable the 'Color Code MIDI Notes By Track' feature (see Figure 4.71).

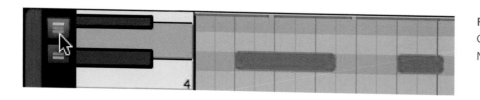

Figure 4.71
Color Code MIDI Notes by Track button.

When this is enabled, tracks in the MIDI Editor are temporarily assigned 1 of 16 fixed colours, in the order they appear in the Track List, as shown in Figure 4.72.

The second button (see Figure 4.73) lets you enable the 'Color Code MIDI notes by Velocity' feature.

Figure 4.72
MIDI Editor colour
coded by Track.

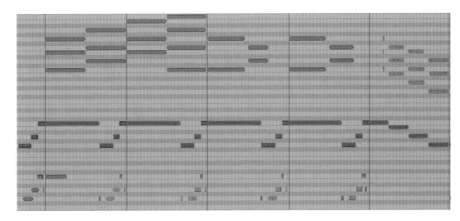

Figure 4.73
Color Code MIDI notes
by Velocity button.

When this is enabled, all the MIDI notes are all displayed in red, but the colour saturation varies from light red for low velocities to dark red for high velocities, based on the Note On velocities – as in Figure 4.74.

Figure 4.74
Colour coded by
velocity.

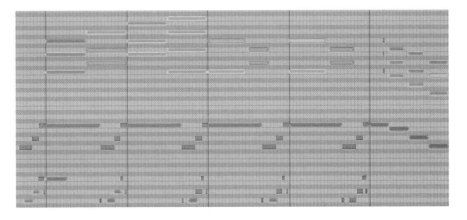

When neither button is enabled, MIDI notes are drawn using the same colours they are in the Edit window, using the colour coding options that you can set in the Track Color Coding settings in the Display Preferences – see Figure 4.75.

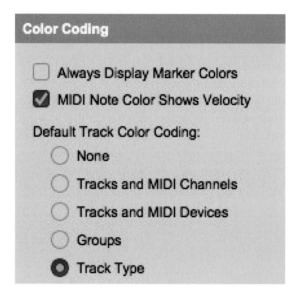

Figure 4.75
Display Preferences:
Track Color Coding
options.

Superimposed Notes View Right-Click Menu

When you are using the Superimposed Notes view, there is a pop-up menu available if you Right-click anywhere in the Notes pane (the part of the window where the notes are displayed) – see Figure 4.76. This provides quick access to the various editing tools, for example, and has an Insert sub-menu that lets you insert key changes, meter changes, and chord symbols at the location of the Edit cursor.

The Separate command lets you separate MIDI notes at the Edit cursor location; the Consolidate command lets you consolidate selected multiple consecutive MIDI notes of the same pitch, and the Mute (or Unmute) Notes command lets you mute (or unmute) selected MIDI notes.

You can also access the MIDI Real-Time Properties and Event Operations, open in a new, un-targeted MIDI Editor window, open in the Score Editor or MIDI Event List windows, or switch to display Notation.

Figure 4.76
The Notes pane
Right-click menu.

Tools	▶
Insert	▶
Cut	
Copy	
Paste	
Merge	
Clear	
Separate	
Consolidate	
Mute	
MIDI Real-Time Properties	
Event Operations	▶
Open in Untargeted MIDI Editor	
Open in Score Editor	
Open in MIDI Event List	
Display Notation	

NOTE

When you are in Notation view, there is an additional 'Notation Display Track Settings' option available in the Right-click menu that lets you open the Notation Display Track Settings dialog and configure the display settings for tracks in Notation view.

The Score Editor

About the Score Editor

The Score Editor in Pro Tools is a great convenience for users who read and write conventional music notation. Just ask the guitar player to read the chords onscreen or print off a copy of the chord sheet, so he or she can learn the part.

Or sketch out a bass part using MIDI, then print this out and ask a bass player to turn this into a 'proper' bass part. Put some strings or brass parts together – you get the idea.

It won't take you long to learn how to use the Score Editor either – it has one main window with a simple tool palette and just a couple of dialog boxes to let you adjust things like the page layout for printing. The tools are the same ones that you will recognize from the Edit window – the Zoomer, the Trimmer, the Selector, the Grabber, and the Pencil – so if you know Pro Tools already, you will know how to use these. Everything is very intuitive. If you see something you want to change, such as the Title or a Clef or a Time Signature, just double-click it and it will usually open up a dialog window to let you change it immediately.

Using the Score Editor

To get started with the Score Editor, you could record some piano or synthesizer pad chords, or import a MIDI file, so that you have some music to work with.

A quick way to get some music into the Score Editor before you have learned how to use its specialized tools is to enter the chords for your music using the Chords ruler.

After you have entered the chords, create a new MIDI track (if there is not one already there) and open the Score Editor. Here you will see the chords laid out above the default music staves. But there will probably be lots of empty bars after the music has finished. To fix this, just click the Double Barline button that you will find among the tools at the top of the Score Editor window and this will create a Double Barline marking at the end of your score – see Figure 4.77 – and the unused bars will disappear from the screen.

If you have used a lot of chords, and especially if you are not a guitarist, you probably won't want to clutter up your score with lots of guitar chord diagrams. To get rid of these, open the Score Setup dialog (see Figure 4.78) by double-clicking on the Title in the Score Editor window. Here you can un-tick Chord

Figure 4.77
Entering a Double Barline at the end of the music.

214

Figure 4.78
Part of the Score
Setup dialog showing
the Information and
Display options.

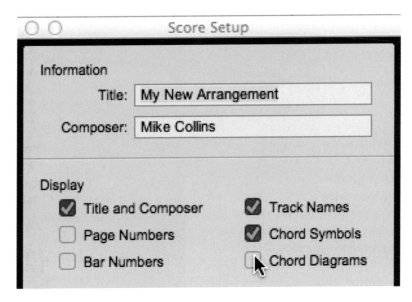

Diagrams in the Display section, so that these will not be displayed. You can also enter the title and the composer credit in the Information section.

At the bottom right of the Score Editor, you will find the Score zoomer arrows – see Figure 4.79. You can use these to size up the score so that you can see just what you need to see in the window.

Figure 4.79
Score Zoomer.

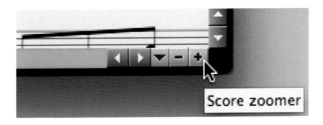

Take a look at Figure 4.80 to see how the Score Editor window might look when you have set everything up this way, with the piano chords visible above the piano stave.

The vertical blue line is called the Cursor Location Indicator. This indicates where the music will start from when you press the Spacebar or click Play on the Transport. If you want to move this, you can either type another location into the Main Counter, or just use the mouse to grab it and move it to wherever you would like to start playback from.

Figure 4.80
Piano chords revealed in the Score Editor.

When you play back the music, a separate Playback Cursor appears in the Score Editor window and moves through the score as the session plays back – see Figure 4.81. This is black, not blue, so you won't get them confused.

Figure 4.81
Playback Cursor visible as a vertical black line to the right of the blue Cursor Location Indicator.

NOTE

You will need plenty of screen space if you want to see lots of the Score Editor onscreen, especially if you are using lots of instruments. A second monitor, or even a 30" monitor, would be good to have. If you are using a smaller monitor, you may not be able to see the whole score, but you can re-size the Score Editor to show just one or two lines of the score and set it to scroll during playback.

Score Editor Pop-up Menu

In the top right-hand corner of the Score Editor window, there is a pop-up menu (see Figure 4.82) that lets you choose whether to show the Selection Display and the Expanded Edit tools in the toolbar that runs along the top of the window.

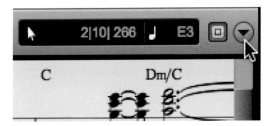

Figure 4.82
Opening the Score Editor pop-up menu.

It also lets you open the Tracks List at the left of the window (which allows you to show or hide your MIDI tracks in the Score Editor) and lets you set the Scrolling Options for the Score Editor window – see Figure 4.83.

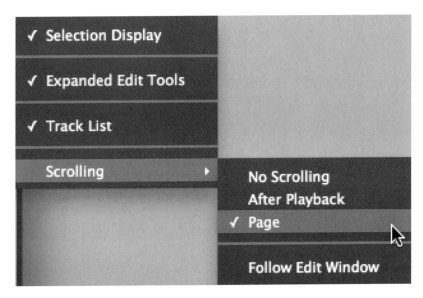

Figure 4.83
The Score Editor Menu.

As you add more MIDI or Instrument tracks, these will show up in the Score Editor using the default piano stave. To change these to single staves, you can either select 'Notation Display Track Settings' from the Tracks List pop-up menu in the Score Editor (see Figure 4.84), or just Double-click a Clef on a Staff.

Figure 4.84
Tracks List pop-up
menu.

The Tracks List pop-up menu (see Figure 4.85) lets you show or hide selected tracks or all tracks, and lets you access the Score Editor's two main dialog windows for 'Notation Display Track Settings' and 'Score Setup'.

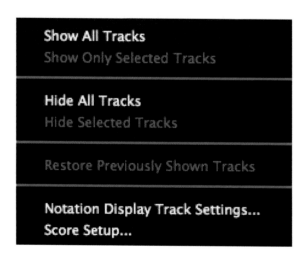

Figure 4.85
Tracks List pop-up
menu.

Setting the Clef

In the 'Notation Display Track Settings' window, you will find a pop-up selector that lets you choose between the Grand Staff that you use for piano, harp, organ, and other keyboard instruments, or the Treble, Bass, Alto, and Tenor clefs used for instruments such as bass, drums, percussion, vocals, strings, or brass – see Figure 4.86.

Figure 4.86
Part of the Notation
Display Track Settings
dialog showing the
clef settings.

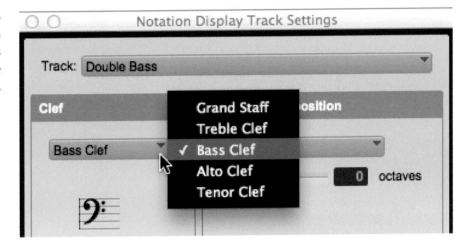

Drum Notation

Music for kit drums is typically written using the bass clef (see Figure 4.87) and the standard five-line musical staff. The hi-hat or cymbal notes are usually written above the top line of the staff, at musical pitch B, written with stemmed X's. You will sometimes see the letters OP to indicate open and CL to indicate closed written over the hi-hat. If there is no indication, the hi-hat remains closed. The small tom-tom is indicated by a note-head placed in the top space, corresponding to the musical pitch G; the snare is in the space below at musical pitch E; the large tom-tom is below this in the space at musical pitch C; with the bass drum in the space below this at musical pitch A. The hi-hat may also be found in the space below the lowest line at musical pitch F, especially if the top line of X's is being used for the ride cymbal. There can be variations to this scheme, with drums written on the lines instead of in the spaces, or with two separate bass clef staves for clarity – usually there will be notes with the music that explain all this.

Hiding Tracks in the Score Editor

To hide MIDI or Instrument tracks that you don't need to see at any particular time, open the Tracks List and click on the small black dot to the left of any track you want to hide – as in Figure 4.88. This will immediately disappear from the Score Editor window.

Using the Notation Tools

To insert a note, you just choose the Pencil tool, select a note duration using the pop-up selector to the right of the Pencil tool (see Figure 4.89), and then click at the pitch and timing location within a bar where you want the note to appear. Rests will automatically be inserted elsewhere in the bar.

Whenever you insert or edit a note in the Score Editor, this note will also appear on (or be changed in) a MIDI or Instrument track in both the Edit and MIDI Editor windows – it's all the same MIDI data.

If you want to hear the MIDI notes play when you click on these in the score, click on the speaker icon to the right of the MIDI Note Duration selector and leave this highlighted. Then when you

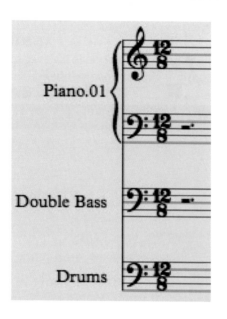

Figure 4.87
Clefs set correctly for piano, Drums, and double Bass.

Figure 4.88
Hiding tracks.

Figure 4.89
Selecting a MIDI Note Duration.

click on any note in the Score Editor (see Figure 4.90), you will hear it play back – assuming that the MIDI or Instrument track that it is based on is set up to play back correctly.

Figure 4.90
Using the Pencil tool
to insert notes.

Trimming Notes

If you want to lengthen or shorten a note, just use the Trimmer tool as you would with a MIDI note or an audio clip – see Figure 4.91.

Figure 4.91
Trimming a note.

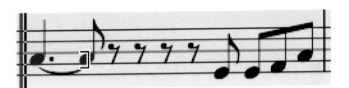

If you point at a selected note on the staff, dragging to the left will make it shorter and dragging to the right will lengthen it – see Figure 4.92.

Figure 4.92
Using the Trimmer
tool to lengthen
a note.

When you get used to working this way, it won't take you too long to enter a complete bass part for a song, as in Figure 4.93.

Figure 4.93
The double Bass part for my new arrangement laid out in the Score Editor.

Copying Notes

Copying notes is easy enough. For example, I decided to copy some of the bass notes in my first arrangement to create a matching drum part. First, I used the Note Selector tool to select the notes in the score, dragging the cursor across the notes to highlight them in blue – see Figure 4.94.

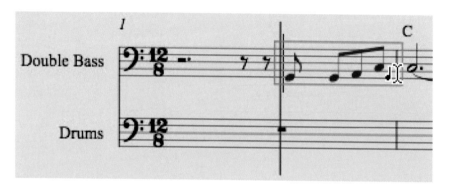

Figure 4.94
Selecting notes using the Note Selector tool.

With the notes selected, I used the standard Copy and Paste commands from the computer keyboard to copy these to an Instrument track that I had set up for the drums – see Figure 4.95.

Figure 4.95
Notes copied to the drums track.

NOTE
Before pasting the notes at a specific location, you need to make sure that the Cursor Location Indicator is positioned at the correct location within the bar. You can just drag the blue Cursor Location Indicator to any location you like within a bar. Alternatively, you can just type the location into the Main Counter.

Moving Notes

You can use the Grabber tool to move selected notes to any other time or pitch location on the same track. With the Grabber tool selected, simply click and drag to move the notes – see Figure 4.96.

Figure 4.96
Moving notes using the Grabber tool.

You can also move notes to another time or pitch location on the same track using the Pencil tool. This changes into a pointing finger that you can use to grab and move notes when you point the mouse at existing notes with the Pencil tool selected – see Figure 4.97.

Figure 4.97
Using the Pencil tool to move a note.

'Transposing Instruments'

A 'transposing instrument' is a musical instrument for which music is written at a different pitch than the pitch that you will hear when the instrument is played. There are two common reasons why this is done: either to make the music easier to read or to make the fingerings on the instrument easier to learn and play. And sometimes both of these reasons apply.

If an instrument has a very high or very low range, the music is typically written either an octave higher or lower than it actually sounds to reduce the use of ledger lines. Examples sounding an octave higher include the celeste and the xylophone, piccolo, and tin whistle. Examples sounding an octave lower include the guitar, bass guitar, double bass, and bass flute. Some instruments with extremely high or low ranges, such as the glockenspiel, use a two-octave transposition. Instruments that 'transpose at the octave' in this way don't play in a different key from concert pitch instruments; instead they just sound an octave higher or lower than written. For example, music for the double bass is written on the bass clef, one octave higher than concert pitch, and music for the guitar is written on the treble clef, one octave higher than concert pitch. Music for the piccolo, on the other hand, is written on the treble clef, one octave lower than concert pitch.

The instruments in families of different sized instruments such as the saxophone, clarinet, and flute have differing ranges of notes that they can play, and these instruments sound lower as they get larger. So that a player can play any member of a particular instrument family using the same fingerings, these instruments are transposed, based on their range, such that each written note is fingered the same way on each instrument. Transposing instruments of this type are often referred to as being in a certain 'key', such as the horn in F. This instrument 'key' determines which pitch will sound when the player plays a note written as C. So, for example, if you play a written C on the French horn (horn in F), it will sound the note F, and if you play a written C on the Bb clarinet you will hear a Bb note. All the other notes on a transposing instrument will, correspondingly, sound at different pitches than written.

Here are some common examples: Instruments in Bb include the Soprano Saxophone, the Bb clarinet, the Trumpet, and the Flugelhorn, and sound a major second lower than written. Instruments in G include the Alto Flute and sound a perfect fourth lower than written. Instruments in F include the Cor Anglais and the (French) Horn and sound a perfect fifth lower than written. Instruments in Eb include the Alto Saxophone and sound a major sixth below what is written. Low bass instruments in Bb include the Tenor Saxophone and the Bb Bass Clarinet (which is sometimes written in treble clef) and sound an octave and a major second below what is written. Bass instruments in Eb include the Baritone Saxophone and sound an octave and a major sixth below what is written. Very low bass instruments in Bb include the Bass Saxophone and sound two octaves and a major second below what is written.

How the Score Editor Handles Transposing Instruments

The Score Editor lets you take account of transposing instruments by allowing you to display notes for any track with a transposition that does not affect the MIDI notes – it just displays the pitches differently in the Score Editor.

You can set these display-only transpositions using the 'Notation Display Track Settings' window – see Figure 4.98. The Key pop-up selector lets you choose the Key Signature that will be displayed and the Octave slider lets you transpose up or down by one or two octaves in either direction.

Figure 4.98
Notation Display
Transposition.

With a Bass Clarinet track recorded using a MIDI instrument to sound at the correct pitch, for instance, you would set the Key pop-up to B-flat and set the Octave slider to +1. Then, if you print out the Bass Clarinet part, a musician will be able to read the part easily and the pitch will still sound correctly – a major ninth below where the part is written.

To get music notation into the Score Editor, you can either write the notes directly into the Score Editor or record them onto a MIDI or Instrument track using step entry or using a MIDI keyboard. You don't have to assign the output of the MIDI track to any instrument if all you want to do is to see the music notation, but most people will probably find it very convenient to play the part using a virtual or real MIDI instrument, so that they can hear how this sounds while they are building up the arrangement.

> **TIP**
>
> If you only want to get notes into the Score Editor so that you can print these out for a musician to read, you might as well go ahead and write these at the transposed pitches in the transposed key. Then, when you print this out, the musician will just play what he sees while the sound from his instrument will be at the correct pitch.

Preparing the Score for a Musician to Read

When you record a musical part with a particular MIDI or virtual instrument, especially if there is an element of 'swing' or a lot of complexity in the rhythms, this can look much too difficult for a musician to read while they are trying to play this.

Composers and arrangers often simplify the music notation to some extent when they are preparing parts for musicians to play. Musicians understand this and will interpret straight notes that are intended to be 'swung' according to the directions of the leader, musical director, arranger, or producer – or according to how they feel the music should be performed. They will make decisions about exactly how long to sustain the notes and what accents to give the notes in the same way – guided by whoever is directing their performance or according to their own experience.

Accordingly, the Score Editor allows you to choose a 'Display Quantization', so that the notes in the score are easier for a human to read, while the underlying MIDI data can use whatever quantization is appropriate so that the sounds produced by MIDI or virtual instruments are correct.

Track Settings

The 'Notation Display Track Settings' window has two tabs that let you set global 'Display Quantization' and other attributes for all the tracks or individual 'Display Quantization' attributes for individual tracks.

Individual Attributes

By default, all the tracks use the global attributes, so if you click on the Attributes tab, you will see that 'Follow Globals' is selected and the various attributes are greyed out.

If you want to set these attributes individually for a particular track, choose the Attributes tab (see Figure 4.99); make sure this track is selected in the Track pop-up at the top of the Notation Display Track Settings window, de-select 'Follow Globals', then go ahead and set the attributes that you want for this track.

Figure 4.99
Notation Display
Attributes set to
'Follow Globals'.

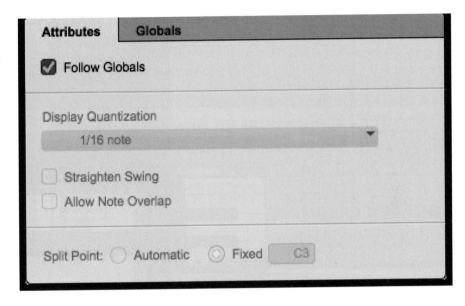

Global Attributes

If you choose the Globals tab (see Figure 4.100), you can set the various attributes that you want to apply to all the tracks that are not individually set up.

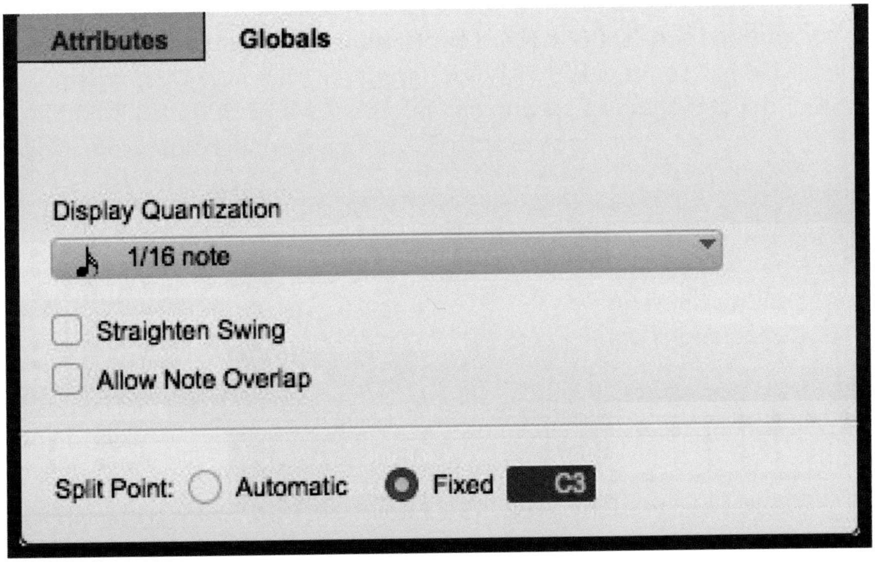

Figure 4.100
Notation Display
Global Attributes.

The Attributes

The main attribute that you will set is the Display Quantization, using the pop-up selector provided – see Figure. 4.101.

If you choose thirty-second note quantization, for example, the thirty-second note subdivisions can be difficult to read at a glance when sight reading – see Figure 4.102.

If you set the Display Quantization to 16th notes, (see Figure 4.103), it doesn't affect the MIDI data in the track, but it does make the music easier to read – especially for a session musician or band member who is feeling the pressure of the 'red light' in the studio.

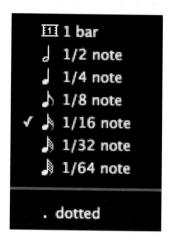

Figure 4.101
Display Quantization
pop-up selector.

Figure 4.102
1/32 note
quantization.

Figure 4.103
Sixteenth note
quantization.

A check box for 'Straighten Swing' lets you set 'swung' notes to be displayed as straight notation. If you select this box, a run of 'swung' eighth notes would be displayed as straight eighth notes, for example.

Sometimes, two notes that start at different times overlap. If this happens, the first note will be truncated when the second note begins. If the 'Allow Note Overlap' check box is selected, the Score Editor will display the full length of any overlapping notes using tied notes.

Setting a Split Point

You can also set a Split Point to divide notes between the treble and bass clefs. When the Clef for the selected track is set to Grand Staff, the selected Split Point setting determines the pitch at which the notes are placed in either the upper or lower staff of the Grand Staff.

If you click in the button marked 'Automatic', the notes will be split between the upper and lower staves of the Grand Staff based on logical note groupings. If you click in the button marked 'Fixed', you can use the associated alphanumeric field to specify a fixed pitch at which to split notes between the upper and lower staves of the Grand Staff.

Printing the Score

When you have prepared all the tracks in the Score Editor to look the way that they should, there are still a couple of things that you may need to attend to in the Score Setup window before you print out the individual tracks or the complete score.

Display Options, Spacing, and Layout

In the Score Setup window (see Figure 4.104) you can choose whether to display the Title and Composer credits, the Track names, the page numbers,

Figure 4.104
Score Setup
window – Information
and Display sections.

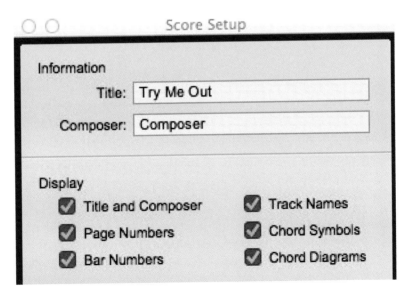

the bar numbers, the chord symbols, or the chord diagrams. As I mentioned earlier, you would not want to clutter up the pages with guitar chord diagrams unless you really do need these. Even more importantly, you can adjust the spacing between the various elements on the page and adjust the layout to make sure that everything is as easy to read as possible.

To open the Score Setup window, either choose Score Setup from the File Menu or select Score Setup from the Tracks menu in the Score Editor window. Or you can simply double-click on the Title.

In the Spacing section (see Figure 4.105), you can enter the spacing you want to use between the staves and between systems of staves. You can also specify the distance to leave below the Title and Composer credits and the first staff of the score and below the chord symbols and diagrams and the top stave of each system.

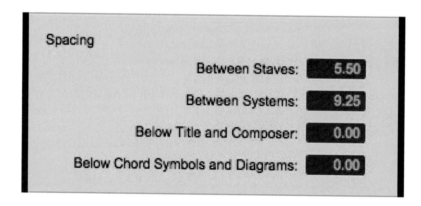

Figure 4.105
Score Setup window – Spacing options.

The Layout section (see Figure 4.106) has a pop-up selector to let you choose the page size: A4, Tabloid, Legal, or Letter. You can also choose whether to use Portrait (tall and not as wide) or Landscape (wide but not as tall) page orientation. You can also specify the Staff Size and the Page Margins in inches or millimetres.

What You See Is What You Get!

When you have everything else set up and ready to print your score, all that is left is to choose whether to print the whole score or individual tracks. What gets printed is what you see in the Score Editor, so all you need to do is to hide any tracks that you don't want to print. If you want to print individual tracks,

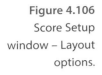

Figure 4.106
Score Setup
window – Layout
options.

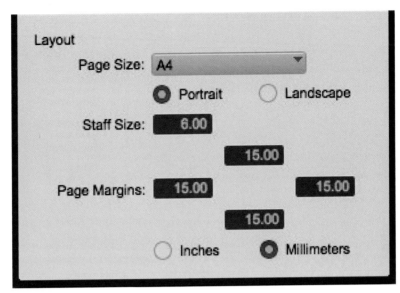

hide all the other tracks and just show one track at a time, print this, then show the next track and print that, and so on.

To print the score, you can either choose 'Print Score' from the File Menu in Pro Tools or press Command-P (Mac) or Control-P (Windows). Your computer's standard Print dialog will appear and you may choose further options here, or just click Print (Mac) or OK (Windows) to print your score.

Exporting to Sibelius

Pro Tools also lets you export the score from your session as a Sibelius (.sib) file. You can then open your score in Sibelius, edit the notation further, and print the score and parts from Sibelius. To export a score from Pro Tools, choose 'Export:To Sibelius' from the File Menu, select a destination and click Save.

You can also send the score directly from your Pro Tools session to Sibelius if you have Sibelius installed on your computer. To do this, you can either Right-click in the Score Editor and choose 'Send to Sibelius' or choose 'Send to Sibelius' from the File Menu. Pro Tools will export all the Instrument and MIDI tracks shown in the Score Editor to Sibelius as a .sib file and will launch Sibelius if it is installed on your computer. As with printing, all you need to do is to hide any tracks that you don't want to export.

Summary

Pro Tools offers a rich set of features for working with MIDI data. If you have MIDI hardware in your studio, such as synthesizers, samplers, drum machines, keyboards or other controllers, or control surfaces to provide hands-on control for virtual instruments and plug-ins, then Pro Tools provides the tools you need to record all your MIDI data. You can choose your method – graphical, Event list or Score – and use the powerful editing features that are provided in the Event Operations and MIDI Real-Time Properties windows.

The separate MIDI Editor and Score Editor windows provide even greater ease-of-use than the main Edit window for many types of editing, but the main Edit window allows you to access everything you need without cluttering up your screen unnecessarily. The MIDI Editor has the advantage that you can view MIDI notes from different tracks superimposed in the same window if you wish – which can be very useful when working on complex musical arrangements. And the Score Editor has the advantage that it is very easy to use and clear to read.

Time spent getting your MIDI equipment organized and set up to work with Pro Tools and preparing templates that can be re-used will be time well spent. If you work out any offsets needed to tighten up the timing aspects of your MIDI gear, this will make a big difference to the results you will be able to achieve.

You should always record MIDI and virtual instruments as audio as soon in your workflow as possible so that you can then reclaim the DSP resources used to create these for use elsewhere in your session. This makes it easier to transfer projects to other software and ensures that you can always hear these parts even if you no longer have access to the hardware or software instruments.

Pro Tools started out as an audio recording, editing, and mixing environment. It now offers an extremely capable MIDI recording, editing, and mixing environment as well.

In this chapter

CHAPTER 5

Recording

Introduction

A complete recording system includes microphones and microphone pre-amplifiers; optional signal processors such as equalizers and compressors; mixing features for multiple microphones; a recording capability with enough channels/tracks; playback facilities for existing music or for a click to play along with; playback facilities for the newly recorded music; a cueing system with headphone amplifiers and headphones for the musicians to hear the click tracks or music playback; and a loudspeaker monitoring system with a talkback system and controls so that the recording engineer, producer, arranger, or others involved in the recording can hear everything they need to and communicate with the musicians.

Pro Tools systems do not include monitor speakers or microphones, and professional Avid Pro Tools hardware interfaces don't incorporate microphone pre-amplifiers – they typically have line level analogue or digital inputs – so you will need to use additional Avid or third-party equipment in most recording situations. Avid does offer the PRE, which is a high-quality eight-channel remotely controllable microphone pre-amplifier unit. However, many recording engineers and producers prefer to use established designs from third-party manufacturers. So most studios will want to add at least a couple of high-quality microphone pre-amplifiers to the equipment list, along with at least one or two (or even a cupboard full of) high-quality studio microphones if they plan to record vocals, guitars, or other acoustic instruments.

Another option, especially for larger studios, is to use a professional mixing console from Neve, SSL, API, or other quality manufacturers. These mixing consoles usually have extremely high-quality microphone pre-amplifiers and plenty of signal processors, such as equalizers or compressors, built-in. Several third-party manufacturers produce mixing consoles that are designed especially to interface with Pro Tools systems, such as the SSL AWS 900 series, which also functions as a Digital Audio Workstation (DAW) controller, or the small format API 1608, which provides 16 vintage-style analogue mic pres and along with 16 equalizers. An external mixer also provides latency-free monitoring while recording; usually has additional connections for playback equipment; and typically provides volume controls, dim, mono and mute switches, talkback features, and switching for control room and studio monitor systems.

Headphone amplifiers with suitable sets of headphones, such as the BeyerDynamic DT100 models, will be needed for the musicians while they record. The recording engineer or producer may be more interested in the super high-quality BeyerDynamic DT48 headphones, which are much more expensive to buy, but which produce sound quality similar to professional studio monitors. For the engineer and producer in the control room, there are plenty of professional studio monitors available from companies such as ATC, PMC and others at the high end, Genelec, KRK, Tannoy, JBL and others in the mid-ground, and more affordable models from Mackie and others at entry level.

Preparing Pro Tools for Recording

If you are just recording one singer or musician, singing or playing without accompaniment and without a click, then all you need to do is to route a microphone through a suitable mic pre via the A/D converters in your Pro Tools interface onto one audio track. Ask the performer to sing or play the loudest that they expect they will during the recording, then set the gain on the microphone pre-amplifier to avoid distortion while keeping the level high enough to maintain a good signal-to-noise ratio. Record-enable the track, press Play and Record in the Transport window, and go ahead with your recording.

While you are learning to use Pro Tools or testing a newly installed system for the first time, it's a good idea to plug a microphone in and try this out first. As soon as you want to get a little more ambitious, recording with more microphones, it will pay you to make a number of other preparations. For example, in the Pro Tools I/O setup window you can enter helpful, descriptive names for the Inputs, Outputs, and Busses. By default, a newly recorded audio file and its clip are named based on the name of the Track onto which it is recorded. So you should also name the track or tracks that you are recording onto with names that will make sense when you look through the recorded 'takes', so that you can more easily identify these later on in the Clips list.

You should make sure that you have set the tempo and meter appropriately and set up a click track for the musicians to play too. If they will be playing along with existing material, you may need to create and adjust one or more headphone mixes and route these into the studio for the musicians. It should never be underestimated how important the headphone mixes can be if you want to get the best results from the musicians. It is always a good idea for the engineer or producer to go out into the studio and listen to the headphones personally to make sure that everything is OK. Often, the musicians may not be able to successfully describe what is wrong with the sound or recognize faults that will be immediately apparent to the engineer or producer.

If you have an external mixer or a DAW controller or monitor controller with talkback facilities, you will be able to talk to the artist from the control room. If not, you should set up a microphone in your control room, route this into an Auxiliary Input in your Pro Tools mixer, and route the output to the musician's headphones – not forgetting to mute this when you are not using it. It can also be a good idea to set up an extra microphone in your recording room, set to omni so it 'listens' to everything, and route this into an Auxiliary channel in your Pro Tools mixer. This way, the musicians can communicate with you even if their instrument or vocal mics are muted.

Disk Space for Recording

Do you have enough disk space to record for as long as you will need to? Pro Tools lets you check how much drive space is available. The Disk Usage window (see Figure 5.1) available from the Window menu, shows the available drive space for each drive connected to your system as text and as a gauge display.

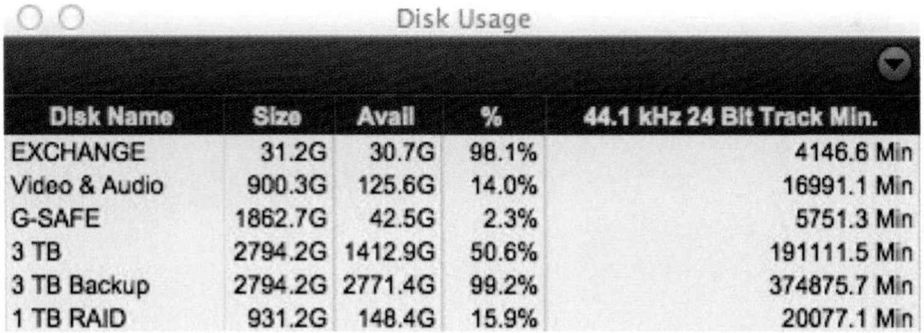

Figure 5.1
Disk Usage window.

Disk Name	Size	Avail	%	44.1 kHz 24 Bit Track Min.
EXCHANGE	31.2G	30.7G	98.1%	4146.6 Min
Video & Audio	900.3G	125.6G	14.0%	16991.1 Min
G-SAFE	1862.7G	42.5G	2.3%	5751.3 Min
3 TB	2794.2G	1412.9G	50.6%	191111.5 Min
3 TB Backup	2794.2G	2771.4G	99.2%	374875.7 Min
1 TB RAID	931.2G	148.4G	15.9%	20077.1 Min

Setting Up Headphone Mixes

In addition to the main mix that you will use to monitor the recordings in the control room, you may also need to create headphone or 'cue' mixes for the musicians and vocalists. The best way to do this is to use Sends to route audio from each track to a pair of available outputs on your audio interface, and route these in turn to a suitable headphone-monitoring amplifier. You can also set up a Master Fader to control the overall level for your cue mix. And if you need different headphone cue mixes, you can use additional Sends, Master Faders, interface outputs, and headphone amplifiers.

To set up a cue mix for all the tracks in your session, you can press Option (Mac) or Alt (Windows) and choose an available Send to insert this on every track. You probably won't want to insert a cue Send on every track, though – there will usually be some tracks that it makes no sense to route into the headphones. So, more usefully, there is a shortcut to insert Sends on selected tracks only: first you select all the tracks on which you wish to insert Sends for your cue mix, then press the Shift key at the same time as holding the Option (Mac) or Alt key (Windows) and choose an available Send.

Figure 5.2
Send window in post-fader mode set to unity gain with Follow Main Pan selected.

TIP
To select a group of tracks, Shift-click on the track names in the Mix or Edit window. To select non-adjacent tracks, hold the Command (Mac) or Control key (Windows) and click on the non-adjacent tracks.

Often, you will be using output paths 1 and 2 from your audio interface to route the main mix from Pro Tools to your monitors. If this is the case, output paths 3 and 4 may be free to use for your cue mix, so you would route the Sends to these outputs. You would then need to route this output pair from your hardware interface into a suitable headphone-monitoring amplifier into which you would plug the headphones.

One way to set up a cue mix is as follows: set each of the Sends to unity gain by Option-clicking (Mac) or Alt-clicking (Windows) on the send fader in each Send window. Also set each Send to post-fader (the default, unhighlighted setting) and enable the Follow Main Pan (FMP) button on each send so that the cue mix will have the same panning as the main mix – see Figure 5.2. This lets you use the same mix for the cue as for your main mix. Of course, you can always raise or lower the level of any individual track in the cue mix using an individual Send fader if the singer or the guitarist wants to hear 'more me'.

Copying Track Automation to Sends

Another way to set up a cue mix is to hold the Option (Mac) or Alt key (Windows) and use the 'Copy to Send' command to copy the current fader settings in the Pro Tools Mix window to the sends. This is a good method to use when you want to set up a headphone mix based on the main mix, or when an effect level needs to follow the levels in a main mix. You can copy the current settings or the entire automation playlist for the selected controls to the corresponding playlist for the send.

You will find this command in the Automation sub-menu in the Edit menu *(Pro Tools HD only)*.

First you need to insert sends on all the tracks that you want to include in your headphone mix, then select the tracks by clicking on the track names, while holding the Shift key to select adjacent tracks or the Command key (Mac) or Control key (Windows) to select tracks that are not next to each other.

When you are ready, select the Copy to Send command from the Edit menu or press Command-Option-H (Mac) or Control-Alt-H (Windows) to open the Copy to Send dialog – see Figure 5.3.

Figure 5.3
Copy to Send dialog.

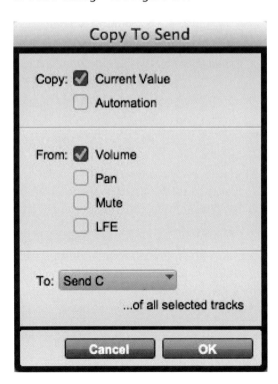

In the Copy to Send dialog, you can choose whether to copy the current values of the volume, pan, mute, or LFE (**Low-frequency effects**, a channel used in surround sound.) controls to the sends, or whether to copy the entire Automation playlist for the corresponding controls.

When you have chosen your settings here, hold the Option key (Mac) or Alt key (Windows) as you click OK to apply the command to all the tracks that you have selected.

Expanded Sends View

If you Command-click (Mac) or Control-click (Windows) on the Send selector for any track, this will change the Sends view to show the individual assignments for level, pan, mute and pre/post fader – see Figure 5.4.

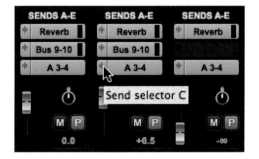

Figure 5.4
Clicking on Send selector C to expand the Sends view. Sends A and B are not expanded. The expanded Sends view is showing the individual assignments for Send C, which is being routed to interface outputs 3 and 4 in this example. Notice that the pre-fader buttons are highlighted in blue to indicate that these are enabled.

You can Option-click (Mac) or Alt-click (Windows) on the Pre/Post button for any Send to set all the Pre–Post buttons to pre-fader (these will be highlighted in blue when selected) so that the Send levels will not be affected by any changes you make to the main mix fader levels. Hold the Shift key while you do this if you just want to apply this change to selected tracks. Using pre-fade sends allows you to tweak the individual send levels to suit the musicians without affecting the main mix.

You can Command-click (Mac) or Control-click (Windows) on any send selector to 'expand' or 'collapse' the send. Similarly, you can Command-Option-click (Mac) or Control-Alt-click (Windows) on any send selector to expand or collapse all sends within the send group (i.e. Sends A–E or Sends F–J). See Figure 5.5 as an example of how the Expanded Sends View looks with several sends expanded.

Figure 5.5
Expanded Sends View
for Sends A, B, and C.

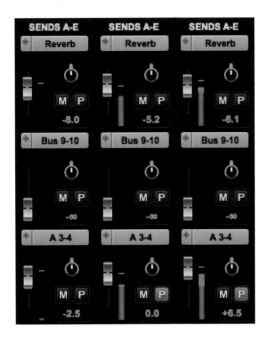

To control the overall level for the cue mix, create a new stereo Master Fader and set this to control the output paths that you are using for the cue mix – in this example, I am using outputs 3 and 4, as can be seen in Figure 5.6.

TIP
If you need additional cue mixes, for the drummer, or for other groups of musicians, you can set up additional sends and route these via additional Master Faders to other available outputs on your audio interface in the same way.

NOTE
If you are working with Delay Compensation enabled, Avid recommends that you do not use any inserts on any Master Fader tracks that you are using to control cue mix levels. Also, you should avoid using inserts on any tracks that you are recording onto, although low latency inserts may not be a problem. Always check that latency is not a problem for the musicians and singers though – and never forget that you are creating the cue mix to help the performers, not for your convenience!

Monitoring Latency

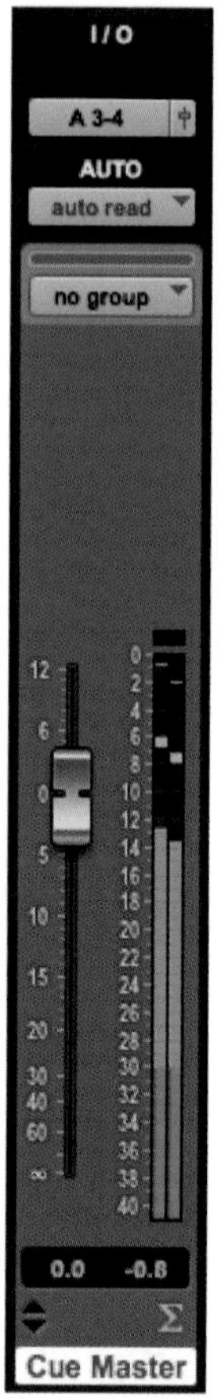

There will always be some delay, typically referred to as *latency*, between the audio coming into Pro Tools and the audio that is sent out from Pro Tools to your monitor speakers. At very least, there will be a delay of a few samples in length due to the analogue-to-digital conversion process on the way in and the digital-to-analogue process on the way out. There may also be other latency delays caused by bussing signals around within the Pro Tools mixer or due to any processing of the audio signals within Pro Tools – which will take at least some small additional amounts of time.

So you will always hear a delayed version of the audio input through your monitor speakers when you are monitoring through Pro Tools. Although you are unlikely to be able to hear the small amounts of latency caused by A/D and D/A conversions, you are more likely to hear delays caused by bussing signals internally or due to processing by plug-ins – and some plug-ins can cause quite long delays that you will have no trouble hearing.

All Pro Tools systems have input-to-output monitoring latency on any record-armed tracks or Auxiliary Inputs with live inputs, and Pro Tools HDX systems have additional monitoring latency on tracks that have one or more host-based Native plug-ins. Native plug-ins use the host processor in your computer, and monitoring latency will occur if you use any Native signal processing plug-ins on the tracks you are recording with. This is particularly noticeable when you are playing Native virtual instruments and monitoring the instrument's audio output, which will suffer from monitoring latency when the host processor receives the MIDI (Musical Instrumental Digital Interface) data and processes this via the plug-in to create the audio. If this delay is clearly audible, this can be very off-putting for a musician who is trying to overdub to an existing performance – because the music already recorded will play back at the correct time while the new performance will be delayed with respect to this due to the monitoring latency.

If you reduce the size of the hardware buffer, this will minimize latency delays. You can do this in the Playback Engine dialog that you can access from the Setup menu – setting the H/W Buffer Size to a smaller number of samples to reduce the latency – see Figure 5.7. But you can never totally eliminate this latency: even at

Figure 5.6
Master Fader set up to control the overall level of a cue mix being sent to interface outputs 3 and 4.

the smallest buffer size, there will still be some latency. Also, reducing the buffer size limits the number of simultaneous audio tracks you can record without encountering performance errors. When you are recording, you will want to set the smallest possible buffer size to counteract latency, but when you have higher track counts with more plug-ins you will need to use larger buffer sizes.

Figure 5.7
Setting the Hardware
Buffer size.

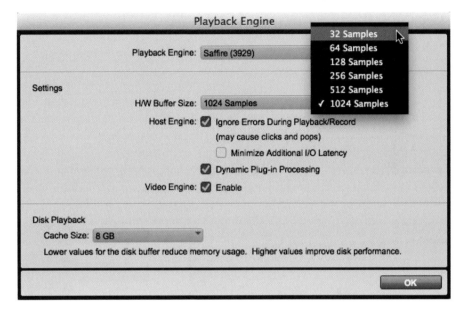

Monitoring via an External Mixer

You can avoid the monitoring latency problem altogether by monitoring the recording source using an external mixer, before it is routed to Pro Tools. Plug your microphones or instruments into the mixer and you will hear the sound directly through a connected monitor system – with no audible delay. At the same time, route the audio to your Pro Tools interface so that you can record it.

> **NOTE**
> Those of you familiar with Pro Tools 10 and earlier versions will recall that it was often necessary to reset the size of the H/W Buffer before recording – and also to disable processor-intensive tracks, particularly those using Virtual Instruments. Pro Tools 11 has a playback buffer that is automatically set for optimum playback and a separate record buffer for record-enabled tracks that users can set to lower values for low latency record monitoring without disrupting the playback buffer. So with Pro Tools 11 you can leave this user buffer set low if you wish – without compromising performance.

> **TIP**
> If you are monitoring through an external mixer while recording, don't forget to defeat the monitoring (pull the fader down or engage the mute button) for each track that you are recording in Pro Tools. Otherwise you will hear a delayed version of the audio mixed in with the undelayed version.

Delay Compensation

When you are using plug-ins or hardware inserts and when you are routing audio within the Pro Tools mixer using sends and busses, delays can occur due to the time needed for the digital signal processing to take place. Pro Tools provides automatic Delay Compensation to compensate for these DSP (Digital Signal Processing) delays. This Delay Compensation automatically compensates for any latencies in the I/O, and maintains phase coherent time alignment between tracks that have plug-ins with differing DSP delays, tracks with different mixing paths, tracks that are split off and recombined within the mixer, and tracks with hardware inserts.

Phase problems have been an issue for Pro Tools users in the past – especially when using plug-ins on some tracks and not on others, or when using different plug-ins with different latency delays, thus introducing phase shifts as a result of the short delays between these tracks. This could be particularly problematic on multimic'ed drum kits – if you put a compressor on the snare, for example, and not on the rest of the kit. Or if complex routings were used with some tracks and not with others – causing delays through the busses or inserts.

Another example would occur if you record a bass guitar to separate channels using a DI (Direct Inject) box for one channel and a microphone in front of a loudspeaker for the other channel. If you then insert a look-ahead peak limiter plug-in (such as Avid's Maxim) on one of these channels, this will delay the output of this track (by 1,024 samples with Maxim) compared with the other track, causing phase cancellations when you mix these together.

DSP plug-ins typically have small processing latencies if they offer static processing such as equalization (EQ), but dynamics processors and some other types may have much longer latency, especially if they use look-ahead techniques. Native plug-ins mostly do not have significant processing latencies, although some do.

It is possible to compensate for these delays by delaying all the undelayed tracks by the same amounts as the delayed tracks. The problem is that this takes time to do and it is all too easy to overlook such details on a busy session. Automatic Delay Compensation solves most of these problems and should always be used while you are playing back and mixing in Pro Tools. It should also be used in most recording situations.

NOTE

When Delay Compensation is enabled, you may have to avoid using inserts on tracks that are being used for recording (or for controlling the levels of cue mixes) because Delay Compensation will create an unwanted time difference between the tracks playing back from disk and any new audio coming into Pro Tools.

A menu item in the Options menu lets you enable or disable Delay Compensation and, to confirm that this is enabled, a Delay Compensation status indicator in the Edit Window Toolbar will appear – see Figure 5.8.

Figure 5.8
The Delay Compensation status indicator is displayed in the Edit Window Toolbar.

TIP

If you have any delays in the session that exceed the maximum available delay compensation, this Delay Compensation indicator will turn red to warn you – so keep an eye out for this! Red can also indicate that you have created a feedback loop that cannot be compensated for. You can hover your mouse over the Delay Compensation indicator in the Edit window to see if the tooltip identifies this as a problem.

When Delay Compensation is enabled, Pro Tools allocates 16,383 samples at 44.1/48 kHz, 32,767 samples at 88.2/96 kHz, or 65,534 samples at 176.4/192 kHz of Delay Compensation for each mixer channel.

> **NOTE**
> Pro Tools adds the exact amount of delay to each track necessary to make that particular track's delay equal to the total System Delay. The total system delay is the longest delay reported on a track, plus any additional delay caused by mixer routing.

The total amount of delay due to inserts and mixer routing for the entire session is displayed in the Session Setup window that you can access from the Setup menu. You can view the reported System Delay in the Session Setup window to check whether or not you are close to exceeding the Delay Compensation limit – see Figure 5.9.

Figure 5.9
The total System delay in samples is displayed in the Session Setup window, accessible from the Setup menu.

In the Operation Preferences page, the Delay Compensation Time Mode setting lets you specify whether Delay values are displayed in samples or milliseconds – see Figure 5.10.

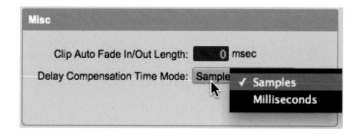

Figure 5.10
Delay Compensation time Mode pop-up selector in the operation preferences.

Delay Compensation View

Although there is a Delay Compensation status indicator in the Edit Window Toolbar that is displayed when Delay Compensation is activated, to see what is going on with the delays on each track you should enable the Delay Compensation View for tracks in the Mix Window.

Figure 5.11
Delay Compensation
View: The structure
instrument track
with sufficient Delay
Compensation
available.

Figure 5.12
Delay Compensation
View: An Aux track
with a reverb
plug-in inserted has
the longest delay in
this session so the
delay parameters are
shown in orange.

Figure 5.13
Delay Compensation
View (in Pro Tools
10): the track on the
left has no inserts;
the track on the right
has a look-ahead
compressor inserted
and the delay values
are shown in red
because there was
insufficient delay
compensation
allocated to
compensate for this.

This view displays the total amount of delay due to DSP and/or Native plug-ins, and any hardware inserts, in the Delay Indicator 'dly' field. You can apply a track delay offset using the User Offset '+/−' field, and you can view the total amount of delay that Pro Tools is applying to the track in the Track Compensation 'cmp' field.

The alphanumeric characters in these fields are displayed in different colours to indicate status. For example, green is used to indicate that everything is working normally and that the track does not exceed the available amount of Delay Compensation – see Figure 5.11.

Orange is used to identify the track reporting the longest plug-in and hardware insert delay in the session – see Figure 5.12.

When the total amount of plug-in and hardware insert delay on a track exceeds the total amount of Delay Compensation available, the alphanumeric characters in the Delay Compensation view turn red, and Delay Compensation is disabled on that track – see Figure 5.13.

If this happens, you can manually compensate for the track delay by moving audio data on the track earlier by the amount of delay reported in the track's Delay indicator.

TIP

In this case, you should bypass the Delay Compensation for the track by Command-Control-clicking (Mac) or Start-Control-clicking (Windows) on the Delay (dly) indicator which will turn grey to indicate that it is bypassed.

Below the Delay Indicator there is an editable numeric User Offset field, identified by '+/−'. While Delay Compensation is enabled, this can be used to type in an offset to the track delay for each track, adding to or subtracting from the amount of delay being applied by the automatic Delay Compensation. You

247

can use this to manually time-align a track if a plug-in is incorrectly reporting its delay, for example, or to adjust the timing 'feel' of the track.

> **TIP**
>
> To bypass the user delay offset, Command-Control-click (Mac) or Start-Control-click (Windows) the User Offset field. With this bypassed, the user-defined delay appears greyed out and does not apply to the track. It can be helpful to bypass the delay offset on a track when you are trying to judge whether to add or subtract a delay, or to decide whether the delay offset is appropriate or not.

The third item, the Track Compensation Indicator, identified by 'cmp', shows the actual amount of Delay Compensation being applied to each track. The colour of the text in the Track Compensation Indicator is shown as follows:

Green indicates that Track compensation is enabled, the track does not exceed the Delay Compensation limit, and the amount of Delay Compensation shown in the indicator is being applied to the track.

Red indicates that the track delay exceeds the available amount of Delay Compensation and no Delay Compensation is being applied to the track.

The Delay Indicator (dly) field will turn Grey when Delay Compensation for the track is bypassed and no delay is being applied to the track – see Figure 5.14.

Figure 5.14
Track Compensation indicator turned grey to show that the Delay Compensation has been bypassed.

> **TIP**
>
> For example, you may choose to bypass Delay Compensation on an Auxiliary Input to let you monitor the audio tracks of a slaved video deck with minimal latency. If you Command-Control-click (Mac) or Start-Control-click (Windows) the Track Compensation indicator, the reported track delay will be zero and will appear greyed out.

Low Latency Monitoring on Audio Tracks

When an Audio track is record-enabled, Track Input Monitoring-enabled, Destructive Punch-enabled, or punched in, the track's Delay Compensation is automatically suspended to provide low latency monitoring on the track outputs. When the track is played back (with both record-enable and TrackInput disabled) it is correctly time-aligned with the other delay-compensated tracks. Tracks that are not record-enabled still apply Delay Compensation. Pro Tools automatically compensates for any timing discrepancies between the recorded material and the delay-compensated mix.

According to Avid Solutions Specialist Simon Sherbourne, 'Where Delay Compensation relates to latency is because it adds a system delay that applies to all tracks. Pro Tools resolves this issue by bypassing Delay Compensation on tracks that are record-enabled or input monitored. Therefore such tracks are *both* utilising the low processing buffer *and* isolated from Delay Compensation to minimise latency. Despite what the manual says, Instrument Tracks and Auxes that contain a virtual instrument with MIDI routed to it always show 0 as their compensation, and MIDI is always kept in time dynamically at the point of playback.

Just because Delay Compensation is suspended on a record track, doesn't mean you'll get low latency. If there are plug-ins with a large inherent delay on the monitoring path (like a look-ahead limiter on the Master fader) you'll still get a delay. For that reason it's good advice to have a cue path free of plug-ins with high latency.

Causing trips between your HDX cards and the host also adds latency to the signal. This happens when mixing Native and DSP plug-ins on a track, or adding Native plug-ins to Auxes or Masters. Again, this is best avoided on a Cue path. The problem is helped in PT11, as any native processors in these configurations are assigned to the low buffer. On native systems, only plug-ins that are downstream of a physical input or VI use the low buffer'.

Overriding Low Latency Monitoring During Recording

You can forcibly apply Delay Compensation, even when the track is record-enabled or Track-Input-enabled, in which case the Track Compensation indicator turns Blue. This is useful for some types of workflow, where overriding Low Latency Monitoring during recording ensures that delay compensation is applied consistently regardless of whether a track is record- or punch-enabled, or whether Pro Tools is recording or playing back.

> **NOTE**
> To forcibly apply Delay Compensation to a track on which Delay Compensation has been suspended, Command-Control-click (Mac) or Start-Control-click (Windows) on the Track Compensation indicator.

Delay Compensation Examples

In the example shown here, a Maxim plug-in is being used on the Pans 3.03 track and this is causing a delay of 1,024 samples. There are no plug-ins on Pans 4, so a delay compensation of 1,024 samples is being automatically applied to this (and to any other tracks in the session that are not causing delays) to keep all the tracks aligned time-wise. There is a ReVibe II plug-in inserted on the Aux track used for Reverb and this is causing a delay of 51 samples. To keep this in step with the other tracks, a compensatory delay of a further 973 samples is being automatically applied – see Figure 5.15.

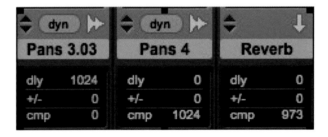

Figure 5.15
The Delay Compensation View with (from the top) the Delay Indicator, User Offset field, and Track Compensation Indicator shown below the mixer tracks.

> **NOTE**
> The Delay Compensation View only shows the insert delay for each track. To view the complete system delay (including any mixer delays), you need to look at the System Delay, which is reported in the Session Setup window – see Figure 5.16. Pro Tools adds the exact amount of delay to each track necessary to make that particular track's delay equal to the total System Delay. The total system delay is the longest delay reported on a track, plus any additional delay caused by mixer routing.

Figure 5.16
The total System
Delay is reported in
the Session Setup
window.

TIP

Don't forget to check the reported System Delay to monitor whether or not you are close to exceeding the Delay Compensation limits: 16,383 samples at 44.1/48 kHz, 32,767 samples at 88.2/96 kHz, or 65,534 samples at 176.4/192 kHz.

NOTE

Pro Tools HD takes converter delays into account when using Automatic Delay Compensation. However, when using non-HD hardware, the System Delay only displays internal delay and does not take into account any latency incurred by the analogue-to-digital (ADC) or digital-to-analogue (DAC) converters in your audio interface.

Mixer Channel Delay Indicator

Each channel strip in the Mix window has a Channel Delay indicator that can be viewed by Command-clicking (Mac) or Control-clicking (Windows) on the numeric Volume indicator located at the left of each channel strip, just below the channel fader – see Figure 5.17. This indicator shows the same insert delay amount for the track that you can also see reported by the Delay Indicator in the Delay Compensation View, so it is useful when the Delay Compensation View is hidden.

Figure 5.17
The Channel Delay
indicator.

> **NOTE**
>
> To the right of the Volume/Channel Delay indicator can be seen the Peak Indicator. This acts as a headroom indicator by displaying the last peak level that has occurred during playback.

Recording with Delay Compensation

Generally, you will want to record with Delay Compensation enabled to maintain phase coherent time alignment between tracks with different DSP delays. However, you should normally avoid using inserts on recording tracks or tracks used for controlling the levels of cue mixes.

Delay Compensation for Side-chains

With Pro Tools HD 11 software, and if you are using Pro Tools HDX hardware, the software will automatically compensate for signal delays in side-chain processing.

A Compensate Side Chains option is provided in the Operation tab of the Preferences window to allow you to disable this if necessary: for example, you may want to disable automatic delay compensation for side-chains with sessions created on previous versions of Pro Tools where you have manually compensated for side-chain delays.

Manual Delay Compensation

If the total delay on a track exceeds the total amount of Delay Compensation available, Delay Compensation will not be automatically applied to that track. As a consequence, any audio on that track will no longer be properly time-aligned and will be out of phase with any other tracks in your session.

Both the Delay indicator and Track Compensation indicator will turn red to show that the track delay has exceeded the Delay Compensation limit and the maximum available Delay Compensation will be applied to all the other tracks.

If this happens, you can manually compensate for the track delays on any tracks that exceed the Delay Compensation limit by moving audio data on these tracks earlier by the amount of delay reported in the track's Delay indicator.

Before manually compensating, you need to bypass the Delay indicator's report to the Pro Tools Delay Compensation Engine. If you Command-Control-click (Mac) or Control-Start-click (Windows) on the Delay indicator, or Right-click the Delay indicator and select Disable Plug-In Delay, the reported delay for the track will be bypassed and the track information will be greyed out. Then all you need to do is to manually nudge any audio data on the track earlier by the amount of delay, which will still be visible in the track's Delay indicator.

> **TIP**
> The Time Adjuster plug-in can be used to manually offset a track in real-time – see Figure 5.18.

Figure 5.18
The Time Adjuster
plug-in.

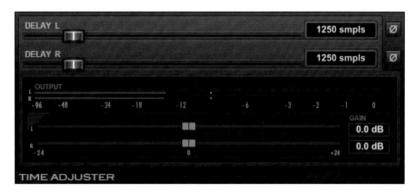

Zero Latency Monitoring

Mbox (third generation), Mbox 2, Mbox Mini, and Mbox Mini 2 all allow you to monitor your analogue input signals while recording – without hearing any latency. Each of these interfaces has a Mix knob on its front panel that you can use to blend and adjust the mix between the interface's analogue input and Pro Tools playback – thus providing 'zero latency' monitoring.

Low Latency Monitoring

Pro Tools HD Native, 003, 003 Rack, 003 Rack+, Digi 002, Digi 002 Rack, Mbox Pro, and Mbox Pro 2 systems each have a Low Latency Monitoring option. This allows you to record to as many tracks as each system supports, with an extremely small amount of monitoring latency. Similarly, certain Core Audio and ASIO audio interfaces that have a built-in mixer (such as the RME Fireface or the MOTU 828) can use the Low Latency Monitoring option.

To use Low Latency Monitoring, make sure that this is selected (ticked) in the Options menu. Only tracks with inputs set to a physical output (not an internal mix bus) use Low Latency Monitoring, so with HD Native systems, you must assign each track to the selected Low Latency Monitoring Path as set in the I/O Setup Output page. For all other systems, you need to assign each track output to either Output 1 or Output 2 for mono output, or to both for stereo output.

> **NOTE**
> When Low Latency Monitoring is enabled, any plug-ins and sends assigned to record-enabled tracks (routed to the selected Low Latency Monitoring Path with HD Native systems, or to Outputs 1–2 with all other systems) are automatically bypassed, and must remain bypassed. Also, these tracks do not register on meters for Master Faders.

Here's how Low Latency Monitoring works during recording: when an audio track is armed for recording (record-enabled), TrackInput-enabled, Destructive-Punch-enabled, or punched in, the track's Delay Compensation is automatically suspended (and the Track Compensation indicator displays 0). This reduces monitoring latency on those track outputs.

When the track is played back (with both record enable and TrackInput disabled), the delay compensation is restored and it is correctly time-aligned with the other delay-compensated tracks.

Tracks that are not record-enabled still apply Delay Compensation. Pro Tools automatically compensates for any timing discrepancies between the recorded material and the delay-compensated mix.

Native Plug-ins 'Health' Warning for Pro Tools|HDX Systems

Using Native plug-ins with Avid HDX systems can seriously deplete the number of available voices in your Pro Tools HD system. They may also add to the latency in your system, so use with care! With Pro Tools|HDX systems, when you initially insert a Native plug-in on an Auxiliary Input or

Master Fader, or on an Instrument track that does not contain an instrument plug-in, and whenever you insert a Native plug-in after a DSP plug-in on any kind of track, this will use two additional voices per channel – one for input and one for output. And if the track is stereo, each channel will use two voices!

The bad news is that Pro Tools also uses another voice for the track if you use the external key side-chain of a Native plug-in on that track, or if you select multiple track outputs for the track (it uses one additional voice for each additional output). And one voice is used for each channel when you select an AFL/PFL (After-Fade Listen/Pre-Fade Listen) Path output in the Output tab of the I/O Setup dialog.

The good news is that adding further Native plug-ins on the same track does not use any additional voices – unless, that is, a DSP plug-in is inserted between two of these other Native plug-ins! Obviously, this is something you should, and can, avoid. Just remember that inserting DSP plug-ins between Native plug-ins on any kind of track will always cause unnecessary voice usage – and may cause additional latency.

So how does Pro Tools handle things when you mix DSP and Native plug-ins? Well, this is a little more complicated to explain. It depends on what mix of DSP and Native plug-ins you are using. Combining Native and DSP plug-ins on an Audio track, Auxiliary Input, or Master Fader, produces different results depending on the order in which they are inserted.

If the Native plug-ins are grouped together and inserted *before* the DSP plug-ins, no additional voices will be used and no processing latency is added. With this configuration, the Native plug-ins will be bypassed when Record Enable or TrackInput monitoring is enabled for that track.

If the Native plug-ins are grouped together and inserted *after* the DSP plug-ins, each initial insert of a Native plug-in after a DSP plug-in on a track will cause processing latency, and will use voices as previously described. And, in this case, Native plug-ins will *not* be bypassed when Record Enable or TrackInput monitoring is enabled for that track.

As a consequence, there are two rules that you should observe when using combinations of DSP and Native plug-ins:

1. Always group plug-ins of the same type together and insert these before any DSP plug-ins to avoid additional voices being used and to minimize processing latency.

2. If you need to make sure that the Native plug-ins stay active when you record-enable a track or use TrackInput monitoring, then you must insert any DSP plug-ins before you insert the Native plug-ins.

Auto Input and Input Only Monitoring

Before you start recording, you also need to consider the monitoring mode and how to set the monitoring levels in the Pro Tools mixer. There are two different ways in which the input signals can be monitored while playing back, while recording, or with the transport stopped. These are the Auto Input and Input Only monitoring modes.

Let's look at the default Auto Input mode in Pro Tools software first. When playback is stopped, the track monitors whatever is coming into its audio input. Now let's consider what happens when you are overdubbing and reach a punch-in point. In this case, you will normally want to hear whatever has been recorded up to the punch-in point. During the punch-in, you want to hear the new audio that is coming in. When you reach the punch-out point, you want the monitoring to switch back so that you can hear the existing track material again. So this is exactly what Auto Input mode does.

Sometimes you don't want to hear the original material that was recorded, you want to hear the incoming audio at all times. This is what Input Only Monitoring does.

These input monitoring options will be very familiar to experienced recording engineers who are accustomed to having input switching available on digital or analogue multitrack recorders. One thing to watch out for here is that Pro Tools does not instantaneously switch to playing back the audio from the track or tracks when you punch out of Auto Input mode – but this is not much of a problem in practice.

When you have *record-enabled* any tracks, a menu command becomes active in the Track menu that allows you to 'Set Record Tracks to Input Only' or to 'Set Record Tracks to Auto Input' depending on which monitoring mode you are currently using. If you prefer to use a keyboard command, you can toggle between Input Only and Auto Input monitoring by pressing Option-K (Mac) or Alt-K (Windows).

With Pro Tools, so that you can tell which mode you are in at a glance, the Input Monitor Enabled Status indicator in the Transport window lights green when Input Only mode is enabled – see Figure 5.19.

Figure 5.19
The cursor arrow is
pointing to the input
Monitor Enabled
status indicator.

With Pro Tools HD, tracks are in Auto Input mode by default and each mixer channel strip has a Track Input button that lets you switch individual audio tracks between Auto Input and Input Only monitoring modes. Again, the Input Monitor indicator lights green when one or more tracks have a TrackInput button enabled – see Figure 5.20.

You can switch modes at any time, during playback, recording, while stopped, and even when a track is not record-enabled. To toggle all tracks in the session, Option-click (Mac) or Alt-click (Windows) any TrackInput Monitor button, and you can apply this command solely to selected tracks in the session if you hold down the Shift key as well. And if you are working in the Edit window, you can enable the TrackInput Monitor button for any track containing the Edit cursor or an Edit selection by pressing Shift-I.

When you have *record-enabled* any tracks, the menu command 'Set Record Tracks to Auto Input' becomes active in the Track menu. This lets you toggle the track input status of all record-enabled tracks between Auto Input and Input Only.

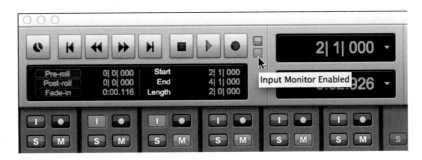

Figure 5.20
When Input Monitor
is enabled on a track,
the Input Status
LED in the Transport
window and the
TrackInput Monitor
buttons for each
enabled track turn
green.

Setting Monitor Levels for Record and Playback

When you record-enable audio tracks, their volume faders turn bright red. This is to remind you that different fader levels can be set for any audio track

> **TIP**
>
> If you have Input Only and Record Enable both selected on a track, Input Only monitoring will be disabled if you deselect Record Enable for the track. Sometimes you will not want this to happen, for example, if you are recording on a series of tracks, one at a time. To prevent this happening, you can disable this behaviour in the Operation Preferences by deselecting the 'Disable 'Input' When Disarming Track (in 'Stop') preference.

during recording and playback. If you *deselect* 'Link Record And Play Faders' in the Operation page of the Preferences window, you can set monitoring levels independently when Pro Tools is recording and when Pro Tools is playing back, for each audio track, and Pro Tools will remember the two different fader levels: one for when the track is record-enabled, and one for when it is *not* record-enabled. When the Operation preference for 'Link Record and Play Faders' is *selected*, the monitoring levels will be kept the same when Pro Tools is recording and playing back, keeping the mix consistent whether you are recording or listening.

Recording With or Without Effects

You need to think carefully about whether to record any effects that you are adding to disk, or whether to use these effects purely for monitoring – perhaps to 'vibe up' the performers while they are playing. Singers usually like to hear at least a little reverb, and guitar players often want to hear distortion or delay effects, for example. One problem with monitoring, but not recording, effects is that the musician will often play things that only make sense when hearing this particular effect and if you don't record this at the same time, you may never be able to recreate this again: you may not be able to get hold of the plug-in or the hardware that was used.

If you want to record the effects that you are adding to voices or instruments to disk, you have more choices to make. You may be adding the effects by bussing signals to and from external hardware. Or you may be routing the audio from the microphone pre-amplifiers into 'outboard' effects before routing this into Pro Tools. It is also possible to use software plug-ins to create effects while you are recording.

Recording With Plug-in Effects

If you insert a plug-in on an Audio track and record audio into this track, the effect is not recorded to disk, it is only applied to the audio coming back from disk. If you want to record the effect to disk, you need to use an Auxiliary Input track. Connect your audio source to this input, insert one or more plug-in effects on the Auxiliary Input, and route the output to the input of an Audio track as in Figure 5.21. Then you can use the Audio track to record the incoming audio together with the sound of the plug-in effects.

Figure 5.21
Recording with plug-in effects inserted on the Auxiliary Input. The output of this channel is routed to the input of an Audio track, which is used to record the incoming audio with the plug-in effects applied to this.

> **NOTE**
> The additional processing via the plug-in adds a further latency delay that can be considerable with certain plug-ins. This is why it is more usual to apply plug-in effects to your recordings after you have recorded the basic tracks.

Recording Audio

When you have set up the tracks you want to record onto, routed microphones into these, set the levels on the microphone pre-amplifiers, set up headphone mixes that the musicians are happy to work with, checked playback of any existing material, sorted out clicks, tempo, meter, key, chords, or whatever, and, ideally, set up Markers for the sections of the music, set up any effects processing (either for the musicians to hear to 'give them the vibe' or that will also be recorded to disk), and set your monitoring levels for the control room and for the headphone mixes, then you are close to getting started! But there are still more moves to make…

Record Enabling

Each track in Pro Tools has a Record Enable button that must be selected so that it lights up red before you put Pro Tools into Record mode using the Transport controls. You can record simultaneously to multiple tracks by record-enabling your choice of Audio, Instrument, or MIDI tracks. When you have clicked the record-enable button for the first track you want to set up, you can add more choices by Shift-clicking on more track record-enable buttons as you decide which you want to use.

> **TIP**
> If you enable the 'Latch Record Enable Buttons' preference that you will find in the Operations section of the Preferences window, then you don't need to Shift-click the Record Enable buttons to enable additional tracks – just click each in turn that you want to be enabled. They will all stay enabled until deselected. And if you want to record-enable all the tracks in your session, if you are recording a band, for example, then Option-clicking (Macintosh) or Alt-clicking (Windows) will do this.

Pro Tools also provides keyboard shortcuts to let you record-enable, input-enable, solo. or mute any tracks that contain the Edit cursor or an Edit selection without using the mouse: simply hold down the Shift key and press R for Record, I for Input enable, S for Solo, or M for Mute to toggle these on or off. If you are working in the Edit window, it can be quicker to use these commands to input-enable and record-enable tracks.

When you have record-enabled all the tracks you want to record onto, click the Record button in the Transport to arm recording, then click the Play button in the Transport to start recording. Alternatively, you can arm and start recording immediately by pressing 3 on the numeric keypad (when the Numeric Keypad mode is set to Transport) or by pressing the function key F12. Or you can simply press Command-Spacebar (Mac) or Control-Spacebar (Windows) – which I always use.

> **NOTE**
> On Mac systems, to use F12 or Command-Spacebar for recording, the Mac 'Spotlight' feature must be disabled or remapped.

If the take gets messed up for any reason, and you realize this before you stop recording, the fastest way to abort the recording is to press Command-Period (Mac) or Control-Period (Windows). This instantly aborts the recording without saving the file – saving you time if you realize that you have a useless take.

If you stop the recording, and then realize that it is no good, you can use the standard keyboard command, Command-Z (Mac) or Control-Z (Windows) to undo your last action – in this case, the recording that you just made.

> **TIP**
> If you are recording a large number of tracks, or playing back a large number of tracks while recording, Pro Tools may take a little longer to begin recording. To avoid this delay, you can put Pro Tools in 'Prime for Record' mode first. Click Record in the Transport so the Record button flashes, then Option-click (Mac) or Alt-click (Windows) on the Play button. The Stop button lights and both the Play and Record buttons flash to indicate that 'Prime for Record' mode is enabled. Now, when you click Play, Pro Tools will begin recording instantaneously.

Other Record Modes

Pro Tools has several other record modes that you can select using the Options menu: Destructive Record, Loop Record, QuickPunch, TrackPunch, and Destructive Punch. You can switch between these by Control-clicking (Mac) or Right-clicking (Windows and Mac) the Record Enable button in the Transport window. When you cycle through these modes, the Record Enable button changes to indicate the currently selected mode, adding a 'D' to indicate Destructive, a loop symbol to indicate Loop Record, a 'P' to indicate QuickPunch, a 'T' to indicate TrackPunch, and 'DP' to indicate Destructive Punch.

Destructive Record mode works like a conventional tape recorder where new recordings onto a particular track replace any previous recordings to that track. There is little justification for using this mode unless you are running out of hard disk space to record to.

Loop Record mode lets you record multiple takes into the same track over a selected time range. Each successive take will appear in the Clips List and can be placed in the track using the Takes List pop-up menu. This is very useful when a singer or musician is trying to perfect a difficult section.

QuickPunch lets you manually punch in and out of record on record-enabled audio tracks during playback by clicking the Record button in the Transport window. This is useful when the engineer or producer wants to decide 'on-the-fly' which bit of a performance to replace while the singer or musician plays throughout the session.

TrackPunch mode lets you manually punch single tracks in and out, or take tracks out of record-enable, without interrupting online recording and playback. You can also simultaneously punch multiple tracks in and out.

DestructivePunch is a *destructive* recording mode that lets you manually punch tracks in and out during playback, replacing the audio material within any track that you punch into and thereby destroying what was there previously. Like TrackPunch mode, DestructivePunch mode lets you punch tracks in and out individually, or punch multiple tracks in and out simultaneously, or take the tracks out of record-enable, without interrupting online recording and playback.

Record Safe Mode

If you are worrying that you might accidentally put the wrong track or tracks into record and mess up a previous recording, then be assured that it is very unlikely that this could happen with Pro Tools. Any new recordings are made to new files by default, so the previous files will still be there in your Clips list and in your session's audio folder. This is what is referred to as 'non-destructive' recording. Nevertheless, Pro Tools does have a 'destructive' recording mode in which new recordings replace previous recordings in the same files.

So, to prevent accidents, Pro Tools provides a Record Safe mode for each track that prevents tracks from being record-enabled. Simply Command-click (Macintosh) or Control-click (Windows) the track's Record Enable button and this will become greyed out and won't let you enable the track to record. If you change your mind, just do this again to get out of Record Safe mode. If you hold the Option key (Macintosh) or Alt (Windows) at the same time, all tracks will be affected. And if you hold the Shift key as well, just the currently selected tracks will be affected.

Half-Speed Recording and Playback

A trick that recording engineers have been using for at least 50 years is to record a difficult-to-play a musical part at half speed an octave below. When played back at normal speed, the part plays back at the correct tempo, but pitched up an octave – back to where it should be.

I remember being introduced to this technique in the early 1980s while recording to analogue tape. I was struggling to play a tricky Clavinet part and the recording engineer just ran the tape at half speed and told me to play along an octave below. When he played what I had recorded back at normal speed it sounded perfect – much tighter timing wise than I had actually played it even at half tempo. I subsequently learned that George Martin used this technique extensively when recording with The Beatles!

Now you can use this technique when recording with Pro Tools. Press Command-Shift-Spacebar (Mac) or Control-Shift-Spacebar (Windows) when you start recording and Pro Tools will play back existing tracks and record incoming audio at half speed.

TIP
If you just want to play back a Pro Tools session at half speed, all you need to do is press Shift-Spacebar. This can be very useful when playing along with or transcribing what is being played on existing recordings – which many musicians need to do from time-to-time.

Recording New Takes Using Playlists

Playlists provide a simple way to be able to keep recording new takes into the same track without any fuss. You can create a new playlist for the track each time you want to record a new take. This saves you the trouble of inserting or choosing a new track and then having to set the track up with the correct inputs and outputs, headphone mixes for the musicians, inserts or plug-ins. Working on the same track you just recorded onto with a new Playlist leaves the previous take just as it was – but hidden from view and disabled from playing – so you can simply record another take into this new Playlist. Every track lets you create as many of these edit playlists as you like, so if you want to record another take, you just create another new playlist.

Each track in the Edit window has a pop-up Playlist selector next to the track name. Click and hold this to reveal the Playlist menu – see Figure 5.22. You can select 'New…' to create a new Playlist. The 'Duplicate' command is useful when you start editing your playlists later on, as is the 'Delete Unused…' Command. All the Playlists you have created for this track are also listed here, and you can even access playlists for other tracks.

Figure 5.22
Pop-up Playlist
selector.

When you select 'New…', you are presented with a dialog box that lets you type a name for the new Playlist – see Figure 5.23 – or you can just OK this and go with the default name if this is fine for you.

Figure 5.23
Naming a new Playlist.

Having created a new Playlist, the first thing you will notice is that the track is now empty – there are no clips visible and the previous take does not play back. So you can go ahead and Record-enable the track, ready to record the next take.

> **TIP**
> Playlists are also very useful for managing differently edited versions of a particular recording. You can chop up a recorded file one way in one playlist and another way in another playlist and then swap these whenever you like while working on your arrangement.

Automatic Punch-in and -out

Because Pro Tools records non-destructively by default, you don't have to take care to punch in and punch out exactly at the right places to avoid recording over material before and after the punch locations, unless you are using a Destructive Record mode. All you need to do is to start recording anywhere before the section you want to replace and stop recording anywhere after the section you want to replace. Then you simply trim the newly recorded and previously recorded clips so that you replace the section of interest from the original recording.

Nevertheless, you can always drop in (punch in) on a track by specifying a range to record to first and setting up a Pre-roll and Post-roll (in the Transport window) as you would do with a conventional multitrack recorder – see Figure 5.24. Playback will start at the pre-roll time and Pro Tools will drop into the record at the punch-in point and drop out of record at the punch-out point, stopping at the end of the post-roll time. Engineers more used to using multitrack tape recorders will probably feel more comfortable with this way of working. And if you really want to 'relive' the experience of working with tape recorders, you can enable Destructive Record to permanently replace the previously recorded audio.

The simplest way to set up a range to record is to select a range in a track's playlist or in a Timebase Ruler at the top of the Edit window using the Selection tool – making sure that the Edit and Timeline selections are linked. As usual, there are other ways to do this – such as typing the start and end times into the Transport window. Set Pre- and Post-roll times if you wish, then go ahead and record.

If you are recording in the default non-destructive mode, a new audio file is written to your hard drive and a new audio clip appears in the record track and Clips List. If you are recording in Destructive Record mode, the new audio overwrites the previous material in the existing audio file and clip.

Figure 5.24
Automatic Punch-in and -out with Pre- and Post-roll enabled.

Pre- and Post-roll

Pre- and Post-roll amounts can be entered in the Transport window or set from a track's playlist or Timebase ruler. Small yellow flags appear in the Main Time Scale ruler to indicate the amounts of Pre- and Post-roll that you have set. When Pre- and Post-roll are enabled, the flags are yellow, otherwise they are white. See Figure 5.25 for an example of how this might look.

Figure 5.25
A linked Timeline and
Edit selection with
Pre- and Post-roll
flags set half a bar
before and after the
selection.

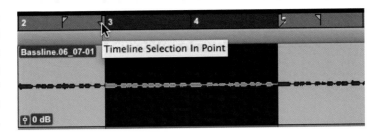

> **TIP**
> You can always set Pre- and Post-roll amounts, store them in a Memory
> Location, and recall the Memory Location whenever you like.

To set and enable the Pre- and Post-roll directly in a playlist, first make sure
that 'Link Timeline and Edit Selection' is enabled, then, with the Selector tool
enabled, drag the cursor across the range that you want to select in the track's
playlist.

> **NOTE**
> When you make an Edit selection with Timeline and Edit Selection linked,
> a pair of Timeline Selection Start and End markers will appear in the
> Timebase ruler. You can drag either of these to lengthen or shorten your
> selection. And if you Option-drag (Mac) or Alt-drag (Windows), you can
> move the selection backwards or forwards along the timeline.

With the Selector tool still enabled, Option-click (Mac) or Alt-click (Windows)
in the track's playlist before the selection to enable the Pre-roll and after the
selection to enable the Post-roll at those locations – see the example in Figure
5.26. Yellow flags will appear in the Timebase ruler. Use Slip mode to allow you
to access any location, or use Grid mode to constrain your choice of location
to grid lines.

If you click on the Pre- and Post-roll buttons in the Transport window to
deselect these, but leave the Pre-and Post-roll amounts set, the flags in the
Timebase ruler will turn white to indicate this – as can be seen in Figure 5.27.

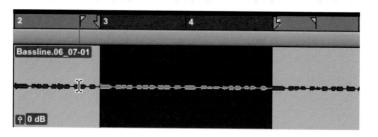

Figure 5.26
Pre-roll location moved closer to the Timeline Selection in-point by Alt-clicking on the playlist using the Selector tool.

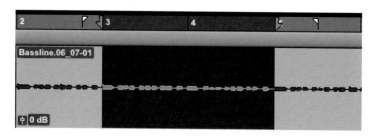

Figure 5.27
White flags indicate that Pre- and Post-roll amounts are set, but disabled.

You can conveniently deselect Pre- and Post-roll in the Transport window and simultaneously remove the Pre- and Post-roll amounts by simply Option-clicking (Mac) or Alt-clicking (Windows) in a playlist near the start of the Edit selection, with the Selector tool enabled.

Alternatively, you can set the Pre- and Post-roll amounts to zero by dragging the Pre- and Post-roll flags until they coincide exactly with the Timeline Selection Start and End markers – see Figure 5.28.

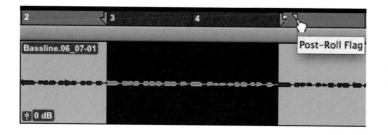

Figure 5.28
Drag the Pre-or Post-roll flags along the Timeline to set or remove the amounts.

The Pre- and Post-roll flags can be moved whenever you like by dragging these in the Main Timebase ruler, either separately or at the same time, and if you set the Edit mode to Grid, the Pre- and Post-roll flags will snap to the current Grid value when you drag them along the Timeline.

To set Pre- and Post-roll values to the same amount, Option-drag (Mac) or Alt-drag (Windows) either the Pre- or the Post-roll flag in the ruler – see Figure 5.29. The other flag will immediately be set to the same value, and will adjust to the identical amount as you drag the selected flag.

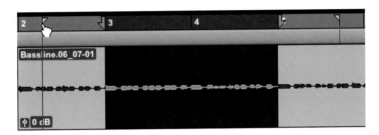

Figure 5.29
Option-drag the Pre- or Post-roll flags to set identical amounts.

Destructive Record

If you prefer to have Pro Tools behave more like a conventional multitrack recorder, where you typically record additional takes over previous takes, you can always enable the Destructive Record mode from the Operations menu. The letter 'D' will appear in the Record button on the Transport to remind you that you are using this mode. When you have recorded the first take, simply leave the track record-enabled, go back and record again. Your new recording will have replaced whatever you had previously recorded in that audio file on disk.

NOTE
If you insert the cursor at the end of your previous recording, the additional material will be appended at the end of the file – thus extending the length of the track.

Now that we have all got used to the idea of Pro Tools hard-disk recording being non-destructive (as long as you always record new takes to new files on disk – which is the default situation), it may seem a little strange for some of you to use the Destructive Record mode. Experienced recording engineers used to working with conventional multitrack recorders will, of course, be perfectly familiar with this way of working. Don't forget, this can be a more efficient way to work – especially if you know what you are doing and are confident that you are not too likely to make a mistake. You won't have to take the time and trouble to sort through the alternate takes and delete them from your hard drive afterwards – and if you are running low on hard disk space Destructive Record can be a boon.

Loop Recording Audio

If you are trying to pin down that perfect eight-bar instrumental solo in the middle of your song or the definitive lead vocal on the verse, it's time to use the Loop Recording feature. To set this up is very straightforward. Just select Loop Record from the Operations menu and you will see a loop symbol appearing on the Record button in the Transport window. Make sure you have Link Edit and Timeline Selection checked in the Operations menu and then use the Selector tool to drag over the clip you want to work with in the Edit window. You can set a pre-roll time if you like, or you can simply select a little extra at the beginning and then trim this back later. Now, when you hit record, Pro Tools will loop around this selection, recording each take as an individual clip within one long file. When you have finished recording, you can choose the best take at your leisure. All the takes will be placed into the Audio Clips list and numbered sequentially, with the last one left in the track for you. Now if you want to hear any of the other takes, just select the last take with the Grabber tool and Command-drag (Macintosh) or Control-drag (Windows) whichever take you fancy from the Audio Clips list and it will automatically replace the selected take in the track – very convenient! An even faster way is to Command-click (Mac) or Control-click (Windows) with the Selector tool at the exact start of the loop or punch range. This immediately brings up the Takes List pop-up – making it even easier to select alternate takes.

But what if you want audition takes from a previous session? These will not be listed here normally, as the start times are likely to be different. The User Time Stamp for each take in Loop record is set to the same start time – at the beginning of the loop – and the Takes List is based on matching start times. That is why it displays all your takes when you bring the List up at the start time of your loop. So if you want to include other takes from a previous session in the Takes List pop-up for a particular location, you can simply set the User

Time Stamp for these clips to the same as for these new takes and they will all appear in the Takes List. And if you plan on recording some more takes later on for this same section, you should store your loop record selection as a Memory Location. This way, these takes will also appear in the Takes List pop-up for that location.

You can also restrict what appears in the Takes List according to the Editing Preferences you choose. If you enable 'Take Clip Name(s) That Match Track Names' then the list will only include clips that take their name from track/playlist. This can be useful when sorting through many different takes from other sessions, for example. Or maybe you want to restrict the Takes List to clips with exactly the same length as the current selection. In this case, make sure that you have enabled 'Take Clip Lengths That Match' in the Editing Preferences.

If you have both of these preferences selected you can even work with multiple tracks to replace all takes on these simultaneously. When you choose a clip from the Takes List in one of the tracks, not only will the selected clip be replaced in that track, but also the same take numbers will be placed in the other selected tracks.

A third option is provided in the Editing Preferences to make any 'Separate Clip' commands you apply to a particular clip apply simultaneously to all other related takes – that is, takes with the same User Time Stamp. You could use this to separate out a particular phrase that you want to compare with different takes, for example.

NOTE

All the clips in your session with the same User Time Stamp will be affected unless you keep one or both of the other two options selected – in which case the Separate Clip command will only apply to clips that also match these criteria.

TIP

It can be easy to forget that you have left this preference selected and end up accidentally separating clips when you don't intend to – so make sure you deselect this each time after using it.

Using Loop Recording

First, make sure that Loop Record is selected from the Options menu so that the loop symbol is shown in the Transport window's Record Enable button. If you are going to set a record or play range by selecting within a playlist, make sure that the Edit and Timeline selections are linked. (Select Link Timeline and Edit Selection from the Options menu or make sure that the Link Timeline and Edit Selection button above the rulers is highlighted.) Choose the Selector Tool and drag the cursor across the range of audio in the playlist that you want to loop record over. Alternatively, you can set Start and End times for the loop in the Transport window. The Loop Record selection must be at least 1 second long.

> **TIP**
>
> Although you can set a pre-roll time that will be used on the first pass and a post-roll time that will be used on the last pass, I recommend that you simply select a loop range that includes some time before and some time after the range you wish to record over. Later, you can trim back the recorded takes to the proper length with the Trim tool.

Record-enable the track or tracks you want to record onto, click Record and Play in the Transport window when you are ready to begin recording – see Figure 5.30 for an example of how this setup might look. Click Stop in the Transport window when you have finished.

Figure 5.30
Setup for Loop Recording.

All the takes are recorded into a single audio file with sequentially numbered clips defined for each take. The most recently recorded take is left in the record track. All of these takes appear as clips in the Clips List and you can audition takes on their own, from the Clips List, or from the Takes List pop-up menu.

Selecting Takes After Loop Recording

Each take created during a Loop Record session is given the same User Time Stamp and you can always access these from the Clips List. To select a different take from the Clips List, use the Time Grabber tool to select the current take in the Edit window then Command-drag (Macintosh) or Control-drag (Windows) another take from the Clips List into the playlist. This will immediately replace the previous take, snapping exactly to the correct location because it has the same User Time Stamp.

Alternatively, you can use the Takes List pop-up menu to select a different take: make sure that the Selector tool is engaged, select the take currently residing in the track, then Command-click (Mac) or Control-click (Windows) anywhere in the clip. The pop-up menu that appears (see Figure 5.31) contains a list of clips that share the same User Time Stamp. So when you choose a clip from the Takes List pop-up menu, this replaces the previous take and snaps exactly to the correct location in the same way.

Figure 5.31
The Takes List pop-up
menu.

> **TIP**
> If you intend to use the track compositing feature after your Loop Recording session, you should enable the 'Automatically Create New Playlists when Loop Recording' preference which you will find in the Record section of the Operation Preferences window. This copies alternate takes created during loop recording to new playlists in the track.

Punch Recording Modes

Pro Tools provides three different manual punch recording modes that all have their uses when you are tracking and overdubbing music.

QuickPunch is a nondestructive punch-recording mode that lets record-enabled tracks be punched in and out during playback by clicking the Record button in the Transport, and provides instantaneous monitor switching on punch-out.

TrackPunch (Pro Tools HD only) is a nondestructive punch-recording mode that lets individual tracks be punched in, punched out, and taken out of record enable without interrupting online recording and playback. TrackPunch also provides instantaneous monitor switching on punch-out.

DestructivePunch (Pro Tools HD only) is a destructive punch-recording mode that permanently replaces pre-existing track material with newly recorded audio in the original audio file. Pro Tools only records while tracks are punched in, inserting new material and replacing audio in the track playlist. As with TrackPunch mode, individual tracks can be punched in, punched out, and taken out of record enable without interrupting online recording and playback.

With both TrackPunch and DestructivePunch, you can record-enable tracks on-the-fly; punch tracks in and out using on-screen record-enable buttons, or remotely from a synchronizer, or from a control surface, or using a foot switch; and compare and match levels using TrackInput switching.

Voice Requirements for Punch Recording

With Pro Tools systems, QuickPunch capabilities depend on the total number of voices available. This will vary according to the number of tracks and plug-ins in use in the current session (which affects the number of available voices). Also, because QuickPunch uses CPU processing, this may reduce the

Preparing for TrackPunch

You can enable TrackPunch mode by selecting this in the Options menu or by Right-clicking the Record button in the Transport window and selecting TrackPunch from the pop-up menu. If you prefer a keyboard command, you can use Command-Shift-T (Mac) or Control-Shift-T (Windows). Possibly the fastest way is to Control-click (Mac) or Start-click (Windows) on the Record button in the Transport window to cycle through available Record modes until TrackPunch mode is selected. When TrackPunch mode is enabled, a 'T' appears in the Record button in the Transport.

Before each pass, you must TrackPunch-enable all the tracks that you intend to punch. You can enable tracks for TrackPunch without record enabling them. This lets you punch in on individual tracks at any time after starting playback by clicking their respective Record Enable buttons. Or, you can simultaneously TrackPunch-enable tracks and record-enable them. This lets recording begin on all the enabled tracks as soon as the transport is record-armed and playback begins.

To TrackPunch-enable a single audio track: Control-click (Mac) or Start-click (Windows) on the track's Record Enable button to toggle the button to solid blue. To TrackPunch-enable or disable *all* audio tracks: Option-Control-click (Mac) or Alt-Start-click (Windows) on any track's Record Enable button to toggle all the Record Enable buttons to solid blue. To TrackPunch-enable or disable all *selected* audio tracks: Option-Control-Shift-click (Mac) or Alt-Start-Shift-click (Windows) on any selected track's Record Enable button to toggle the Record Enable button for the selected audio tracks to solid blue. These actions TrackPunch-enable the tracks, but do not record-enable the tracks, so you will still need to click on the individual tracks that you wish to punch in on.

Or, you can simultaneously TrackPunch-enable tracks *and* record-enable them. This lets recording begin on all the enabled tracks as soon as the transport is record-armed and playback begins.

To simultaneously TrackPunch-enable and record-enable a single audio track: Click the track's Record Enable button so that it flashes from blue to red. To simultaneously TrackPunch-enable and record-enable all audio tracks: Option-click (Mac) or Alt-click (Windows) any track's Record Enable button so that all tracks' Record Enable buttons flash from blue to red. To simultaneously TrackPunch-enable and record-enable all selected audio tracks: Option-Shift-click (Mac) or Alt-Shift-click (Windows) any selected track's Record Enable button so that the Record Enable button for the selected audio tracks flash from blue to red.

After putting the Transport Record button into TrackPunch mode, when you first click on any individual track record enable button, this will both TrackPunch-enable it and record-enable the track. To confirm this, the small Record Enable Status LED to the right of the Transport Record button goes red – see Figure 5.34. Also, when a track is both TrackPunch-enabled and record-enabled, the track record-enable button will flash from blue to red.

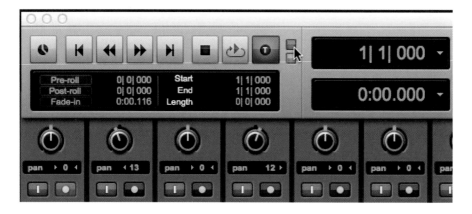

Figure 5.34
When a track is TrackPunch-enabled, and is also record-enabled, its Record Enable button lights flashes from blue to red, the Transport window Record button also flashes blue to red and has a 'T' in the middle, and the small Record LED to the right of this is lit.

If you click once more on the track's record-enable button, this will disable it for recording, but will leave it TrackPunch-enabled, and the Record Enable Status LED in the Transport window will be unlit. With TrackPunch enabled (but not record-enabled) for a track, the Track record-enable button will light up solid blue, as in Figure 5.35.

To disable a track that you have TrackPunch-enabled, so that it is neither TrackPunch-enabled nor record-enabled, Control-click (Mac) or Start-click (Windows) on the track record-enable button. This will no longer be lit and, when no tracks are TrackPunch- or record-enabled, neither will the Record button in the Transport window.

Figure 5.35
When a track is
TrackPunch-enabled,
but not record-
enabled, its Record
Enable button lights
solid blue and the
Transport window
Record button has a
'T' in the middle, but
the small Record LED
to the right of this is
not lit.

NOTE
If at least one track is TrackPunch-enabled, the Record button in the Transport window lights solid blue. When TrackPunch mode is enabled but no tracks are TrackPunch-enabled, the Transport Record button flashes from grey to red. If at least one track is TrackPunch-enabled, the Transport Record button flashes from blue to red. If at least one TrackPunch-enabled track is also record-enabled, the Transport Record button flashes from blue to red, *and* the small record LED to the right of this also lights up red. Whenever at least one audio track is recorded, the Transport Record button lights up solid red.

Using TrackPunch

When you are all set up to use TrackPunch, you can either punch in on individual tracks or punch in on multiple tracks simultaneously when you reach the first punch-in point during playback. You may also want to punch in immediately, then punch out further into the track, then punch in again later on.

To punch in on individual tracks, make sure that each track you want to punch in on is TrackPunch-enabled, with its Record Enable button lit solid blue. Click Record in the Transport to enter the TrackPunch Record Ready mode, with the Record button flashing from blue to red. Then click Play in the Transport to begin playback. While the session plays back, you can punch in and out on individual TrackPunch-enabled tracks by clicking their record-enable buttons.

> **TIP**
>
> After a TrackPunch recording pass, the punched track's playlist in the Edit window displays the clips created by punching. You can use any of the Trim tools after punch recording to open up the head or tail of TrackPunch recorded clips, or to reveal the parent audio file that was recorded in the background. This lets you compensate for any late or missed punches.

To punch in on multiple tracks simultaneously, make sure that each track you want to punch in on is both TrackPunch-enabled and Record-enabled. To punch in or out on all the TrackPunch-enabled tracks simultaneously, play back the session and click the Record button in the Transport window whenever you like.

> **NOTE**
>
> Alternatively, click Record in the Transport first, then Option-Shift-click (Mac) or Alt-Shift-click (Windows) on any selected track's Record Enable button to simultaneously punch in and out on all the currently selected TrackPunch-enabled tracks.

To start your TrackPunch pass in the record, punching in immediately on all tracks: make sure that all the tracks are both TrackPunch-enabled and Record-enabled and that the Transport is in the TrackPunch Record Ready mode, then click Play on the Transport to begin playback. During playback, you can punch out and then punch in and out again whenever you like on any TrackPunch-enabled track by clicking its Record Enable button.

DestructivePunch

DestructivePunch, only available for Pro Tools HD, is a destructive version of TrackPunch mode. Where TrackPunch always records audio to a new file,

DestructivePunch destructively records audio directly into the original file, using a fixed 10-millisecond linear crossfade. No additional clips are created when recording in DestructivePunch mode. Up to 200 'running punches' can be performed on a track during a single DestructivePunch pass.

Preparing for DestructivePunch

To use DestructivePunch on an audio track, the track must contain a continuous audio file of a minimum length, which you can set in the Pro Tools Operation preferences page. Also, the file must start at the beginning of the session (at sample 0) and the File Length must be equal to or greater than the DestructivePunch File Length setting. If the DestructivePunch File Length is not sufficiently long for the files you are working with, it is easy enough to change this in the Operation Preferences so that it is equal to or greater than the length of the current file – see Figure 5.36.

Figure 5.36
Setting the
DestructivePunch
File Length in the
operation preferences.

Sometimes you may want to use DestructivePunch with a file that does not start at the beginning of the session, and that you cannot move to the start of the session because this would put it out of sync with other files. And very often a playlist will contain clips taken from different files. In these cases you can either use the Bounce to Disk command or use the Consolidate command to create a continuous file of the required length that starts at the beginning of the session.

Because this is such a common situation, Pro Tools has a special 'Prepare DPE Tracks' command in the Options menu that lets you consolidate audio on all DestructivePunch-enabled tracks. To use this command, you need to enable DestructivePunch mode first, then make sure that all the tracks you want to prepare are DestructivePunch-enabled, but not record-enabled. When you use the 'Prepare DPE Tracks' command, Pro Tools consolidates audio on all DestructivePunch-enabled tracks from the beginning of the session to the value specified in the DestructivePunch File Length preference.

You can enable DestructivePunch mode by selecting this in the Options menu or by Right-clicking the Record button in the Transport window and selecting

> **NOTE**
> For clips with any clip gain settings other than 0 dB, using the Prepare DPE
> Tracks command in DestructivePunch mode automatically renders all clip
> gain settings and resets all clip gain settings to 0 dB.

DestructivePunch from the pop-up menu. A faster way is to Control-click
(Mac) or Start-click (Windows) on the Record button in the Transport window
to cycle through available Record modes until DestructivePunch mode is
selected. When DestructivePunch mode is enabled, 'DP' appears in the Record
button in the Transport.

Before each pass, you must DestructivePunch-enable all tracks that you
intend to punch. You can enable tracks for DestructivePunch without record
enabling them. This lets you punch in on individual tracks at any time after
starting playback by clicking their respective Record Enable buttons. You may
also need to do this if you want to use the Prepare DPE Tracks command to
consolidate files for Destructive-Punch recording.

To DestructivePunch-enable a single audio track: Control-click (Mac) or
Start-click (Windows) the track's Record Enable button to toggle the button
to solid blue. To DestructivePunch-enable or disable *all* audio tracks: Option-
Control-click (Mac) or Alt-Start-click (Windows) any track's Record Enable
button to toggle all Record Enable buttons to solid blue. To DestructivePunch-
enable or disable all *selected* audio tracks: Option-Control-Shift-click (Mac) or
Alt-Start-Shift-click (Windows) a selected track's Record Enable button to
toggle the Record Enable buttons for the selected audio tracks solid blue.
These actions DestructivePunch-enable the tracks, but do not record-enable
the tracks, so you will still need to click on the individual tracks that you wish
to punch in on.

Or, you can simultaneously DestructivePunch-enable tracks *and* record-
enable them. This lets recording begin on all the enabled tracks as soon as the
transport is record-armed and playback begins.

To simultaneously DestructivePunch-enable and record-enable a single
audio track: Click the track's Record Enable button so that it flashes from
blue to red. To simultaneously DestructivePunch-enable and record-enable
all audio tracks: Option-click (Mac) or Alt-click (Windows) a track's Record
Enable button so that all tracks' Record Enable buttons flash from blue to red.
To simultaneously DestructivePunch-enable and record-enable all selected

audio tracks: Option-Shift-click (Mac) or Alt-Shift-click (Windows) any selected track's Record Enable button so that the Record Enable buttons for the selected audio tracks flash from blue to red.

After putting the Transport Record button into DestructivePunch mode, when you first click on any individual track's Record Enable button, this will both DestructivePunch-enable it and record-enable the track. To confirm this, the small Record Enable Status LED to the right of the Transport Record button goes red – see Figure 5.37. Also, when a track is both DestructivePunch-enabled and Record-enabled, the track's Record Enable button will flash from blue to red – Figure 5.38 shows an example of how this might look.

Figure 5.37
Record Enable
Status LED.

If you click once more on the track's record-enable button, this will disable it for recording but will leave it DestructivePunch-enabled, and the Record Enable Status LED in the Transport window will be unlit.

Figure 5.38
DestructivePunch
is enabled in the
Options menu and
the Record button
has been selected
in the Transport; an
Audio track is also
Record-enabled: so
the Transport Record
button flashes blue
to red, the individual
track record-enable
buttons flash blue
to red, and the small
LED to the right of
the Record button in
the Transport window
lights up red.

With DestructivePunch enabled (but not record-enabled) for a track, the Track's Record Enable button will light up solid blue.

To disable a track that you have DestructivePunch-enabled so that it is neither DestructivePunch-enabled nor record-enabled, Control-click (Mac) or Start-click (Windows) on the track's record-enable button. This will no longer be lit and, when no tracks are DestructivePunch- or record-enabled, neither will the Record button in the Transport window.

> **NOTE**
>
> If at least one track is DestructivePunch-enabled, the Record button in the Transport window lights solid blue. When DestructivePunch mode is enabled but no tracks are DestructivePunch-enabled, the Transport Record button flashes from grey to red. If at least one track is DestructivePunch-enabled, the Transport Record button flashes from blue to red. If at least one DestructivePunch-enabled track is also record-enabled, the Transport Record button flashes from blue to red, *and* the small record LED to the right of this also lights up red. Whenever at least one audio track is recorded, the Transport Record button lights up solid red.

Using DestructivePunch

When you are all set up to use DestructivePunch, you can either punch in on individual tracks or punch in on multiple tracks simultaneously when you reach the first punch-in point during playback. You may also want to punch in immediately, then punch out further into the track, and punch in again later on.

To punch in on individual tracks, make sure that each track you want to punch in on is DestructivePunch-enabled, with its Record Enable button lit solid blue. Click Record in the Transport to enter the DestructivePunch Record Ready mode, with the Record button flashing from blue to red. Then click Play in the Transport to begin playback. While the session plays back, you can punch in and out on individual DestructivePunch-enabled tracks by clicking their record-enable buttons.

To punch in on multiple tracks simultaneously, make sure that each track you want to punch in on is both DestructivePunch-enabled and Record-enabled.

To punch in or out on all the DestructivePunch-enabled tracks simultaneously, play back the session and click the Record button in the Transport window whenever you like.

To start your DestructivePunch pass in the record, punching-in immediately on all tracks: make sure that all the tracks are both DestructivePunch-enabled and Record-enabled and that the Transport is in the DestructivePunch Record Ready mode, then click Play in the Transport to begin playback. During playback, you can punch out and then punch in and out again whenever you like on any DestructivePunch-enabled track by clicking its Record Enable button.

> **NOTE**
> Alternatively, click Record in the Transport first, then Option-Shift-click (Mac) or Alt-Shift-click (Windows) on any selected track's Record Enable button to simultaneously punch in and out on all the currently selected DestructivePunch-enabled tracks.

Delay Compensation and DestructivePunch

It is important to be aware of whether or not Delay Compensation was in use when the recording that you wish to use DestructivePunch with was originally recorded. If Delay Compensation was used on the original recording and this is not used when you punch in using DestructivePunch, or vice versa, the new audio will not be recorded into exactly the correct position in the audio file. It will be offset in one direction or the other by the amount of the Delay Compensation. To make sure that this does not happen, if Delay Compensation was inactive when recording the original file, it should be deactivated while using DestructivePunch and if it was active when recording the original file, it should be kept active while using DestructivePunch.

Consequently, you may need to manually override the automatic suspension of Delay Compensation that occurs when record-enabled tracks switch to Input Monitoring when you are using DestructivePunch. Take a look at Figure 5.39 and notice how the Delay Compensation has been automatically suspended with Input Monitoring enabled.

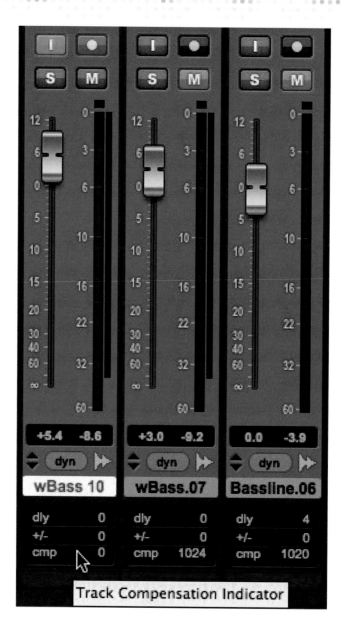

Figure 5.39
Track Compensation
is automatically
suspended by the
Auto Low Latency
feature when Input
Monitoring is enabled.
A Dyn3 Compressor/
limiter being used
as an insert on both
the wBass tracks,
and, although delay
compensation of
1,024 samples is
being applied to
wBass.07, with
Destructive Punch
and Input Monitoring
both enabled on
wBass 10, the Track
Compensation
Indicator shows
that no delay
compensation is
being applied to this.

This feature is called Auto Low Latency and can be turned off by Right-clicking the Track Compensation indicator on the track and selecting Auto Low Latency Off, or by Command-Control-clicking (Mac) or Control-Start-clicking (Windows) on the Track Compensation indicator on the track – see Figure 5.40.

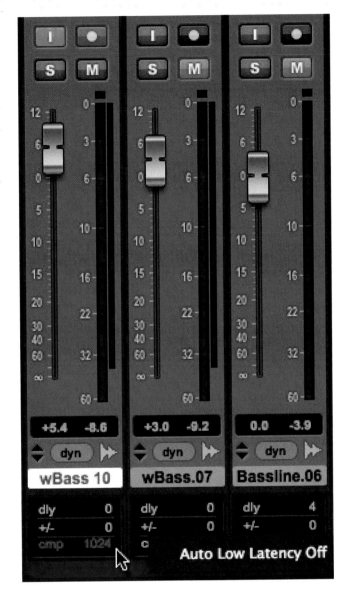

Figure 5.40
Track Compensation
suspension can be
overridden if you
Command-Control-
click (Mac) or
Control-Start-click
(Windows) on the
Track Compensation
indicator.

After Punch Recording

When you are finished with the record pass after Punch recording, the track Record Enable status and transport Record Arm status for the tracks you have been recording onto will follow the current Audio Track RecordLock and

> **TIP**
> To apply Delay Compensation to all selected tracks where Delay Compensation was suspended, either Shift-Right-click the Track Compensation indicator on any selected track and select Auto Low Latency Off, or Command-Control-Shift-click (Mac) or Control-Start-Shift-click (Windows) on the Track Compensation indicator on any selected track.

Transport RecordLock settings. You can set these in the Record section of the Operation Preferences – see Figure 5.41.

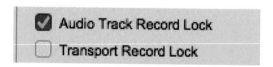

Figure 5.41
RecordLock preferences.

If the Audio Track RecordLock option is enabled, the record-enabled audio tracks remain armed when playback or recording stops. When the Audio Track RecordLock option is not enabled, record-enabled audio tracks are taken out of record enable when Pro Tools is stopped – as with a digital dubber.

If the Transport RecordLock preference setting is not enabled, the Transport Record disarms when Pro Tools is stopped. When enabled, the Transport Record remains armed when playback or recording stops so you don't have to rearm the Transport between takes – as with a digital dubber.

So, for music production, you will normally have the Audio Track RecordLock preference selected to enable it and the Transport RecordLock preference un-ticked to disable it. If you are using Pro Tools as a digital dubber, then you would reverse these preferences.

Summary

If you have followed this chapter thoroughly, you will be able to set up Pro Tools to record in a variety of scenarios. Setting up headphone cue mixes; strategies for combining DSP and Native plug-ins if you are using a Pro Tools HD system; input monitoring; adjusting monitoring levels during record and playback; recording with or without effects; recording through plug-ins – you need to get all of this stuff sorted out before you get started with serious recording projects.

Whether you are overdubbing one musician at a time or recording a whole band, you should know how to arm tracks in the various record modes, how to record or play back at half speed, how to punch in and out automatically or manually, how to loop record audio, and how to use playlists to record alternate takes.

You also need to understand how and where latency delays occur and develop strategies to avoid these delays that take into account the needs of the musicians you are recording and the results that you are aiming to achieve. Get to know how the delay compensation features work and you will be on your way to mastering the Pro Tools recording environment. And to make those quick fixes that are needed with even the most practiced studio musicians, you need to master the various punch modes – choosing just the right length of pre-roll in bars and beats that gives the musicians enough time to prepare for the punch-in, without wasting time by starting too many bars before the punch location.

Pro Tools has all the features you need to record just about anything that makes a sound that can be captured by a microphone, but if you are new to this software environment, and even if you are upgrading from older systems, there is plenty that is new to learn here. It may take a while before you get an opportunity to use all these recording features, so feel free to revisit this chapter whenever you like to remind yourself of how everything works.

In this chapter

CHAPTER 6

Editing

Introduction

To become proficient at editing using Pro Tools, you need a firm understanding of how the various Edit modes work and how to select material to edit using the various editing commands. Understanding what clips are and how to manipulate these is also crucially important. Knowing the best ways to create loops and edit these is particularly important when you are working in the various pop, rap, and dance music genres, where the Dynamic Transport mode can also be very helpful. For example, Track Compositing offers an easy-to-use workflow, but seasoned Pro Tools users may choose to continue using their previously tried and tested methods. As usual, there are several ways of achieving the same results with Pro Tools – it is one of the most flexible environments available for editing audio!

The Edit Menu Commands

You can access most of the commands that you will use for editing from the Edit menu, (see Figure 6.1), but you will also find some commands, such as Capture and Loop, that could be considered as editing commands in the Clip menu and vice versa (I'm thinking about Trim Clip and Separate Clip here). Gaining familiarity with the Edit menu and the editing commands in the Clip menu should be priorities for aspiring Pro Tools editors.

Some of these commands, such as Select All, don't require any additional explanation or are explained perfectly well in the Pro Tools Reference Guide, while explanations of others, such as the Automation editing commands, belong elsewhere. One of the fastest ways to learn the keyboard commands is to look through these menus, note the keyboard equivalents for the various commands, and then use these as often as possible from memory.

Undo and Undo History

Pro Tools tracks the last 64 undoable operations that you carry out, storing these in a queue that you can step back through using the Edit menu's Undo command or by pressing Command-Z (Mac) or Control-Z (Windows). If you

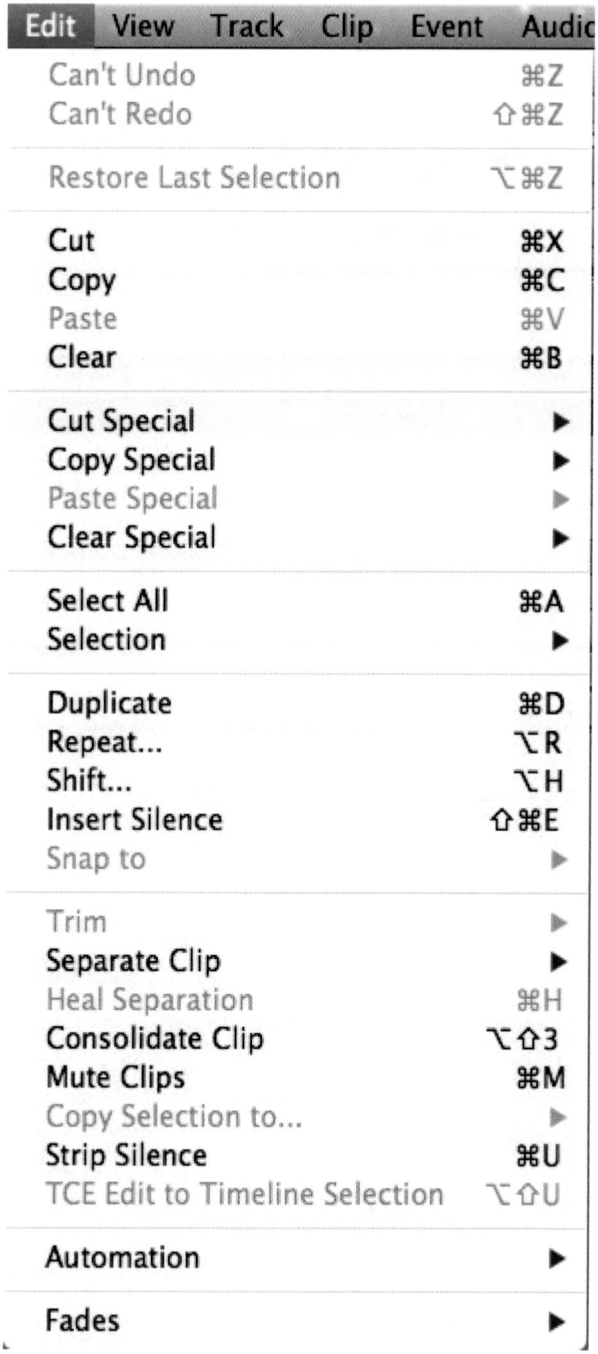

Figure 6.1
Pro Tools HD edit menu.

change your mind about this, you can step forward through the undo queue by choosing the Redo command from the Edit menu or by holding the Shift key while you press Command-Z (Mac) or Control-Z (Windows).

When you reach the 64-step limit and perform another undoable operation, the oldest stored operation is removed from the queue. Keep in mind that it is possible to set a lower limit for the number of undoable steps in the Editing Preferences, so if you are working on someone else's system, or if someone else may have reset this, you should check to make sure what limit is set.

You can also use the Undo History window (see Figure 6.2) that you will find in the Window menu. This lets you view the queue of undoable and redoable operations. It shows undoable operations in bold type, and if you click on any of these, it will undo all the undoable operations back to and including the one you clicked on and will show these as redoable operations in italic type. Click on any redoable operation to redo all operations up to and including the one you clicked on.

The Undo History can optionally show edit creation times, which can be a help when you are deciding which operation to undo or redo. This option can be accessed using the pop-up selector at the top right corner of the Undo History window. You can also undo or redo all operations in the queue or clear the whole queue.

Figure 6.2
Undo History window
showing the creation
times of the edits.

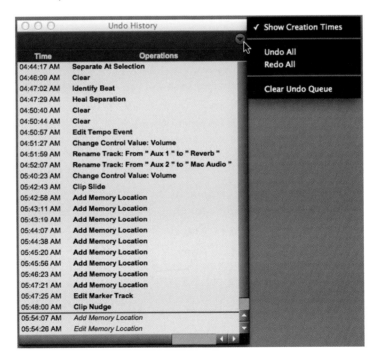

Restore Last Selection

Very often, you will make an Edit or Timeline selection and then lose it. Perhaps you accidentally click somewhere in the Edit window without realizing or whatever. Pro Tools has a really useful command to get you out of trouble here! Just choose 'Restore Last Selection' from the Edit menu or press Command-Option-Z (Mac) or Control-Alt-Z (Windows) and your selection will magically reappear!

Cut, Copy, Paste, and Clear Commands

Pro Tools has some similarities with word-processing and graphics software. For example, the way that the computer expects you to work is to select something first and then to issue a command that says what you want to do with whatever you have selected – either using keyboard commands or using the mouse, pointer, and menus. Once you have selected a clip, for example, you can use the Cut, Copy, Paste, and Clear commands to re-arrange and edit the material in your tracks.

You can select a clip or clips using the Time Grabber tool or you can select a range along a track using the Selector tool. You can also work across multiple tracks. You can then use the Cut command to remove whatever you have selected from the Edit window and put this selection into the Clipboard ready to paste it elsewhere. The Clipboard is a temporary storage area in the computer's RAM (random access memory). You can use the Copy command to copy your selection into the Clipboard ready to paste elsewhere, without removing the original selection from the Edit window. You can use the Paste command to put the contents of the Clipboard into the Edit window at the Edit insertion point, overwriting any material that is already there. If you simply want to remove your selection without putting it into the Clipboard, use the Clear command instead.

Special Cut, Copy, Paste, and Clear Commands

Pro Tools also has four 'special' Edit menu commands that you can use for editing automation playlists (e.g. volume, pan, mute, or plug-in automation) on audio, Auxiliary Input, Master Fader, VCA Master, and Instrument tracks. The Copy and Paste Special commands let you copy or paste automation data into another clip (without affecting associated audio or MIDI notes). They also let you edit automation and MIDI controller data without having to change Track Views.

Cut Special, Copy Special, and Clear Special let you cut, copy, or clear Clip Gain settings, and each has three additional sub-menu selections. These let you edit all the Automation or just the Pan Automation (whether the data is showing in the track or not) or just the Plug-in Automation on its own (when this is showing in the track).

The Paste Special command lets you paste Clip Gain settings and also has three additional sub-menu selections, as follows.

Merge lets you add the pasted data to any existing automation data of the same type in the destination selection. This can be useful for consolidating MIDI data from several tracks into a single MIDI track. For example, you like the pitch bend on the synth trombones and want to apply it to the synth guitar.

To Current Automation Type pastes the automation data or MIDI controller data from the clipboard to the selection as current automation type or MIDI controller. This lets you copy automation from any type to any other type. For example, you can copy Left Pan automation to Plug-In automation. You can also copy MIDI volume data and paste it to MIDI pan.

The Repeat to Fill Selection command allows you to automatically fill a selection with audio or MIDI clips or data much more quickly than by manually duplicating the clips.

To use the 'Repeat to Fill Selection', simply cut or copy an audio clip so that it is in the Pro Tools software's Clipboard (i.e. temporarily stored into RAM), then make a selection in the Edit window, and use the command to fill this selection. The Batch Fades window automatically opens to let you apply crossfades between the pasted clips.

When pasting audio clips, you are prompted to specify a crossfade to be used for the pasted clips. If you fill an area that is an exact multiple of the copied clip size (e.g. filling 16 bars with a 4-bar clip), the copied selection is pasted as many times as it takes to fill the selection.

If you fill an area that is not an exact multiple of the copied clip size (e.g. filling 15 seconds of a track with a 2-second clip of room noise), the remaining selection area is filled with an automatically trimmed version of the original selection.

> **NOTE**
> These commands can also be used for MIDI controller data on MIDI and Instrument tracks.

Cutting, Copying, and Pasting Clip Gain

Pro Tools lets you cut, copy, and paste clip gain settings, so that you can apply the clip gain settings from one clip to any other, although they cannot be cut, copied, or pasted across clip boundaries.

To cut clip gain settings, select a single whole clip or make an Edit selection within a single whole clip and then choose Cut Clip Gain from the Cut Special sub-menu in the Edit menu or Press Control-Shift-X (Mac) or Start-Shift-X (Windows).

> **TIP**
> You can cut clip gain settings from any Edit selection if you just want to get rid of them, but you can only paste the clip gain settings that you have cut from a single clip.

To copy clip gain settings, first select a single whole clip or make an Edit selection within a single whole clip. Then choose Copy Clip Gain from the Cut Special sub-menu or Press Control-Shift-C (Mac) or Start-Shift-C (Windows). Or you can right-click any single whole clip selection and choose Copy Clip Gain from the Clip Gain sub-menu if you prefer.

> **NOTE**
> Clip gain breakpoints on the clipboard are time-stamped with the playback times in the timebase of the track being copied (which means you can cut or copy, then paste clip gain settings from clips on tick-based tracks, and have the pasted clip gain settings match the corresponding bar:beat locations of clips on other tick-based tracks).

To paste clip gain settings, first make an Edit selection within a single whole clip, placing the Edit In Point where you want the cut or copied clip gain settings to be pasted. Then choose Paste Clip Gain from the Edit menu or press Command-V (Mac) or Control-V (Windows).

NOTE

The clip gain settings on the clipboard are pasted into the clip starting at the Edit In Point. The clip gain settings are pasted in their entirety, but only apply to a single clip. If the pasted clip gain settings extend beyond the end of the clip, they are all still associated with the clip. This means that if you trim out the clip later, the pasted clip gain is revealed. When pasting clip gain within a clip (rather than to a single whole clip of the same duration), breakpoints are added before and after the pasted data so that any clip gain settings outside the paste do not change.

Stripping Silence from Clips

The Strip Silence feature (see Figure 6.3) lets you analyze audio selections containing one or more clips across one or more tracks to define areas that you wish to regard as 'silence'. Once you have identified these, you can 'strip away' (i.e. remove) the 'silence' or you can keep (i.e. 'extract') the 'silence' by removing the rest of the audio. The third option is to separate your selection into lots of smaller clips so that these can be quantized or edited individually afterwards.

Four sliding controls in the Strip Silence window let you set the parameters by which 'silence' will be defined.

TIP

If you press Command (Mac) or Control (Windows) while adjusting the sliders you get finer resolution.

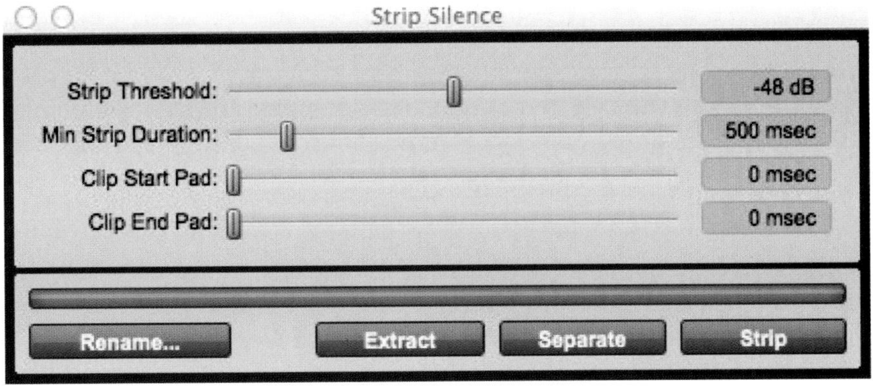

Figure 6.3
The Strip Silence
window.

299

The Strip Threshold parameter lets you set a value for the amplitude, below which any audio is considered to be silence. For example, on a bass drum track, you will probably hear the sound of the rest of the drum kit when the bass drum is not playing, but this will be much quieter than the bass drum – in other words, the amplitude of this quieter audio will be much less than the amplitude of the bass drum sound. You can define this low amplitude audio to be considered as silence by setting the Strip Threshold just above this.

> **NOTE**
> Strip Silence lets you adjust the Strip Threshold down to –96 dB. This extended dynamic range is especially useful when working on recordings with low signal levels (such as ambient recordings) and recordings with a wide dynamic range (such as orchestral music).

You can also define what is to be considered to be silence using the Minimum Strip Duration parameter. This sets the minimum amount of time that the material below the threshold must last for before it is regarded as 'wanted' audio.

The Clip Start Pad parameter lets you extend the start of each new clip to include any wanted audio material that falls below the threshold, such as breathy sounds before a vocal or the sound of fingers sliding up to a note or chord on a guitar. Similarly, the Clip End Pad parameter lets you extend the end of each new clip to make sure that the full decay of the audio material is preserved.

As you adjust the controls, rectangles start to appear in the selected clip surrounding the wanted audio. The audio in between these rectangles is considered to be silence.

For example, take a look at Figure 6.4 on the next page. This shows a four-bar selection from the bass drum track of a multi-mic'ed drumkit. The bass drum inside the selection plays on the two and the four beats (yes, it is reggae). These notes are outlined by rectangles. Everything in between is considered to be silence.

Figure 6.4
With the Strip Silence parameters correctly set, you see black rectangular boundaries around the 'wanted' audio.

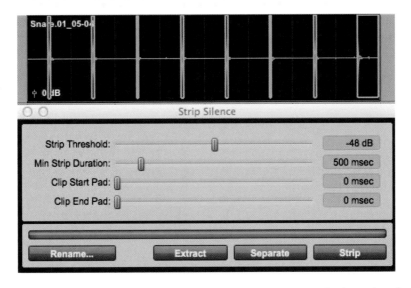

If you pad the start, this opens up the rectangles before the drum beat, and if you pad the end, this opens up the rectangles after the drum beats – including more audio before and after the drum beats. As you can see in Figure 6.5, it is unable to do this before and after the clip selection. To assess how much to pad the sounds, you will need to strip the silence, audition carefully, undo the strip silence and adjust the pad settings, strip the silence again and audition carefully again, and repeat this until you have the results you want. Listen carefully to the decay of the sound, for example. If you cut this too short, it can make the recording sound too unnatural.

Figure 6.5
Adjusting the start and end pads.

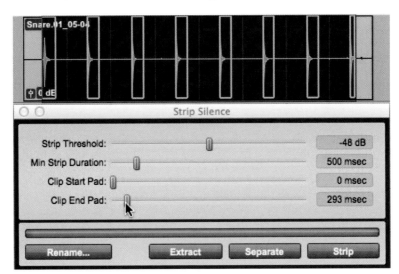

301

When you click the Strip button, it removes the areas that you have defined as silence from the selected clip. In the accompanying screenshot, Figure 6.6, you will see the five 'wanted' clips containing the bass drum notes. Now you can move the bass drums around to change the timing or 'feel' – or edit these whatever way you wish.

Figure 6.6
The 'silence' between the 'wanted' clips has been stripped away by clicking 'Strip'.

If you click the Extract button instead (see Figure 6.7), this removes the audio above the Strip Threshold and leaves (or 'extracts') the audio that you have defined as silence. So you can think of Extract as the inverse of Strip with the 'wanted' audio in this case being the audio that falls below the Strip Threshold. This feature could be useful in post-production if you wanted to extract the 'room tone' or ambience from part of a recording to use elsewhere, for example.

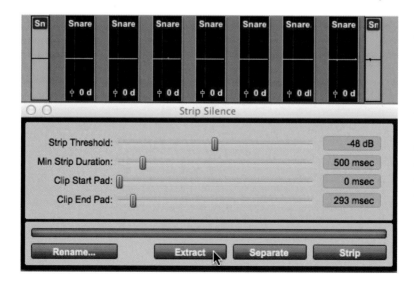

Figure 6.7
The audio above the Strip Threshold has been removed, leaving the audio defined as 'silence' (but which actually contains ambience) remaining (i.e. 'extracted') in the track.

The third option is to click the Separate button. This automatically divides the selection into clips based on the boundaries detected by Strip Silence. In the accompanying screenshot, Figure 6.8, you will see that the selection has been divided into separate clips, each with its own name.

TIP
The Separate feature is very useful if you want to quantize the audio in each clip to line up with the bars and beats, for example.

Figure 6.8
The Separate button divides the whole selection into separate clips based on the boundaries detected by Strip Silence.

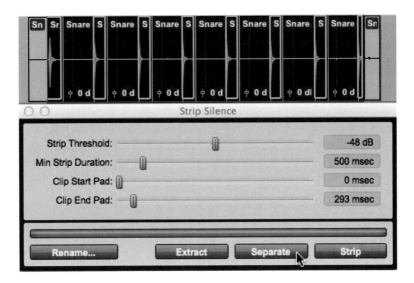

If you click the Rename button in the Strip Silence window, this opens the Rename Selected Clips dialog – see Figure 6.9. This lets you define how the clips that you will create using the Strip Silence feature will be named.

Whatever you type in the Name field will be used as the base name for the clips created using Strip Silence. For example, if you want a space to appear after the name and before the numbers, you will have to insert a space here. The Starting Number field lets you specify the number at which sequential auto-numbering starts. The Number of Places field lets you specify the number of zeroes to use before the clip numbers. You can also specify a Suffix to be used at the end of the name, after the number. You might use this to identify the 'take' from which these clips have been created, for example.

For this example, I chose 'Bass Drum' with a space after it as the Name; Starting Number 1; three as the Number of Places to give me two leading zeroes with each number up to 9, one leading zero before each number from 10 to 99, and no leading zeroes for numbers above 100. For my Suffix, I typed a space, followed by a hyphen, then another space, then 'Take 2' to identify the take from which these bass drums had been stripped.

Figure 6.9
Rename selected clips dialog.

Take a look at the accompanying screenshot, Figure 6.10, to see how the clips were named in the Clips List as a result of this naming scheme.

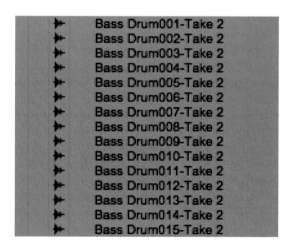

Figure 6.10
Bass Drum clips in Clips List showing the naming scheme.

Inserting Silence

The Insert Silence command does what it says, replacing a selection that you have made on audio, MIDI, or Instrument tracks with silence. The Insert Silence command also affects markers in any selected Conductor Rulers (such as the Tempo or Meter rulers).

Less obviously, it can be used to remove automation data. If the track is displaying audio or MIDI data, when you apply the Insert Silence command to a selected range, it not only clears the audio or MIDI data, but it also clears any automation data for the track or tracks. However, if the selected tracks are displaying automation data, the automation data visible on each track is cleared throughout the selected range and any audio or MIDI is left untouched. Also, if you press the Control key (Mac) or Start key (Windows) while choosing the Insert Silence command, this clears all the automation data from all the selected tracks – not just the visible data.

The Insert Silence command is particularly clever when used in Shuffle mode. In this case, it moves the track data within and after the Edit selection forward, by an amount equal to the selection, pushing everything (including any automation data) forward from the start of the selection – to get it 'out of the way' of, thus making room for, the silence that you are inserting. To explain this another way: when you apply the Insert Silence command to an Edit selection while in Shuffle mode, Pro Tools splits the clip, or clips, at the beginning of the insertion point and moves the new clip, or clips, to a position later in the track by an amount equal to the length of the selection, before inserting the selected amount of silence.

TCE (Time Compression and Expansion) Edit to Timeline Selection

Pro Tools HD software has a very useful 'TCE Edit to Timeline Selection' Edit menu command – see Figure 6.11 – that lets you compress or expand an audio selection to fit the Timeline selection.

Figure 6.11 Edit menu TCE Edit to Timeline Selection command.

To fit an Edit selection to a Timeline selection, first disable 'Link Timeline and Edit Selection', then select the audio material to be compressed or expanded using the Selector tool; select the time range where you want to fit the audio material using any Timebase ruler – see Figure 6.12 – and then choose 'TCE Edit to Timeline Selection' from the Edit menu. If you prefer a keyboard command, you can use Option-Shift-U (Mac) or Alt-Shift-U (Windows).

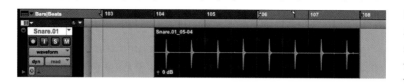

Figure 6.12
A four-bar Edit selection ready to be fitted to a three-bar Timeline selection.

The Edit selection is compressed or expanded to the length of the Timeline selection – see Figure 6.13.

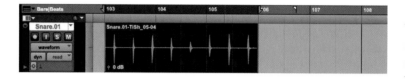

Figure 6.13
The Edit selection has been compressed and moved to fit into the Timeline selection using the 'TCE Edit to Timeline Selection' command.

On Audio tracks, the 'TCE Edit to Timeline Selection' command uses whichever TCE AudioSuite Plug-in you have selected in the Processing Preferences, and on Elastic Audio-enabled tracks, it uses whichever plug-in you have selected for use with Elastic Audio – see Figure 6.14.

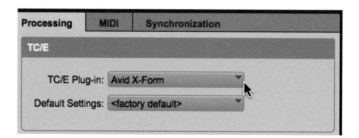

Figure 6.14
Selecting the TC/E Plug-in preference.

TIP

The Pro Tools Time Shift plug-in, although usable to give you a rough idea of how things will sound as you are working ideas out, will not give you the best results. To achieve better results, you will need to buy a much higher-quality plug-in. I can recommend Serato's Pitch 'n' Time, which has very professional features. Avid also offers an excellent time compression and expansion plug-in called X-Form, which I am currently using. This takes quite a lot longer to process the audio, but delivers truly professional results.

> **NOTE**
> The 'TCE Edit to Timeline' command can be used on multi-channel selections and selections across multiple tracks. All clips are compressed or expanded equally by the same percentage value, based on the Edit selection range. This ensures that the rhythmic relationship between the different channels or tracks is maintained.

Consolidating Clips

Often, during your editing sessions, you will end up with tracks made up from many smaller clips. This could happen if you 'comp' several takes to make a composite containing the best from each take. When you are satisfied with your edits to these tracks, Pro Tools allows you to consolidate a track or a range within a track (such as a verse or chorus) into a single clip, which is much easier to work with. When working with audio tracks, this consolidation process causes a new audio file to be created that encompasses the selection range, including any blank space, treating any muted clips as silence.

To use this feature, first you select the clips you want to consolidate using the Time Grabber tool or the Selector tool. A quick way to select all the clips in a track is to triple-click in its playlist with the Selector tool. When you choose Consolidate from the Edit menu, this will create a new, single clip that replaces the previously selected clips. If you prefer to use a keyboard command instead, you can press Option-Shift-3 (Mac) or Alt-Shift-3 (Windows).

> **TIP**
> Consolidating an audio track does not apply any automation data as it creates the new file, so if you want to create a new file with automation data applied to the audio, use the 'Bounce to Disk' feature instead.

The Edit Modes

Pro Tools has four Edit modes – Slip, Shuffle, Spot, and Grid – that can be selected by clicking the corresponding button in the upper left of the Edit window. You can also use the function keys on your computer keyboard, F1 (Shuffle), F2 (Slip), F3 (Spot), and F4 (Grid), to select the mode. The Edit mode that you have chosen affects the ways that clips may be moved or placed, how commands like Copy and Paste work, and how the various Edit tools (Trim, Selector, Grabber, and Pencil) work.

Slip Mode

The default mode that you should normally work in is the Slip mode. It is called Slip mode because it lets you move clips freely within a track or to other tracks and allows you to place a clip so that there is space between it and other clips in a track.

In Slip mode, if you use the Cut command to remove a selection before the end of the last clip on a track, it leaves an empty space where the data was removed from the track. You should also be aware that clips are allowed to overlap or completely cover other clips in Slip mode.

Shuffle Mode

In Shuffle mode, if you place two or more clips into a track, they will automatically snap together with no gap in between. If you have an existing track that contains two or more clips with gaps between them, you can close the gaps by selecting Shuffle mode and using the Time Grabber tool to push a clip in the direction of the previous clip – see Figure 6.15.

Figure 6.15
Selecting and moving a clip in Shuffle mode.

As soon as you let go of the mouse, the gap between the two clips is closed, leaving them 'stuck together' – see Figure 6.16.

So, if you want a clip that you are moving around in the Edit window to automatically butt up against the previous clip, with no overlap and with not

even the smallest gap between them, you can use Shuffle mode. If you are in Slip Mode or even in Grid Mode, you have to be careful to avoid overlapping the clips, and you will have to zoom in to see what you are doing. Consequently, placing clips accurately in these modes takes more time.

By the way, it is called 'Shuffle' mode because if you use the grabber to drag a clip placed earlier in a playlist to a later position (or vice versa), the other clips will shuffle (i.e. move) their positions around to accommodate this re-positioning of the clip. Similarly, if you use the Cut command to remove a selection before the end of the last clip on a track, the clips to the right of the cut move to the left, closing the gap. Also, if you paste data anywhere before the end of the last clip on a track, all clips beyond the insertion point move to the right to make room for the pasted material.

> **TIP**
> Be careful to return to Slip or Grid mode as soon as you have made your moves in Shuffle mode – it is all too easy to accidentally move a clip and have Shuffle mode shuffle your clips to somewhere they shouldn't be. And if you don't notice this at the time it happens, you may not be able to use even the multiple Undo feature to get back to where you were.

Spot Mode

Spot mode was originally designed for working to picture, where you often need to 'spot' a sound effect or a music cue to a particular SMPTE time code location.

> **NOTE**
> Because it can be so important to exclude Shuffle mode to ensure that clips stay time-aligned while editing, a Shuffle Lock feature is available that prevents you from inadvertently entering Shuffle mode by disabling all key commands and control surface switches for Shuffle mode. To enable or disable this, Command-click (Mac) or Control-click (Windows) the Shuffle button on-screen. A lock icon appears in the Shuffle button when this mode is enabled, and you must switch to another mode before you can use this feature.

The way this works is that you select Spot mode and then click on any clip in the Edit window or drag a clip from the Clip List or from a Workspace browser into the Edit Window. The Spot dialog comes up and you can either type in the location you want or use the clip's time stamp locations for spotting.

There are actually two time stamps that are saved with every clip. When you originally record a clip, it is permanently time-stamped relative to the SMPTE start time specified for the session. Each clip can also have a User Time Stamp that can be altered whenever you like using the Time Stamp Selected command in the Clips List pop-up menu. If you have not specifically set a User Time Stamp, the Original Time Stamp location will be set here as the default.

Spot mode can also be very useful when editing music projects in Pro Tools, especially if you move a clip out of place accidentally, which can easily happen in Slip or Shuffle modes. Using Spot Mode, you can always return a clip to the location on the timeline at which it was originally recorded.

Sometimes, of course, you may deliberately move a clip to a location other than where it was first recorded. If you do this, I recommend that you take the trouble to set a User Time Stamp at this new location so that you can always return the clip to this new location if you should accidentally move it.

If you do move a clip by accident (see Figure 6.17), using Spot Mode is an ideal way to put it back to exactly where it came from.

To do this, put the software into Spot mode by clicking on the Spot mode icon at the top left of the Edit window, select the Grabber tool, and click on the clip to bring up the Spot dialog – see Figure 6.18.

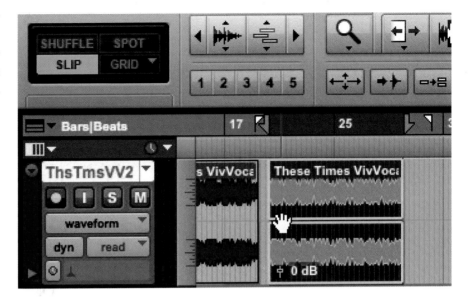

Figure 6.17
Edit window showing
a Clip accidentally
moved from its
original position.

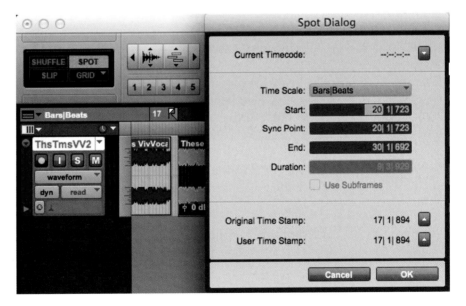

Figure 6.18
Using the Spot Dialog
to return a Clip to its
original position.

If the Clip has been moved from the position at which it was originally recorded, the current Start position displayed in the Spot Dialog will be different from the Original Time Stamp position shown in the dialog. You can either type the correct Start position or (even quicker) just click the upwards-pointing arrow next to the Original Time Stamp to put this value into the Start position field – see Figure 6.19.

311

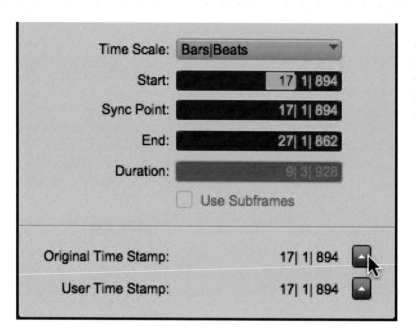

Figure 6.19
Click the Original Time
Stamp arrow to enter
this location into the
Start field in the
Spot Dialog.

NOTE

If you have deliberately moved the Clip since recording it and have taken
the trouble to enter a User Time Stamp in the Clips List for the moved clip,
you should enter the User Time Stamp into the Start field instead. If you
did not update the User Time Stamp when you moved the clip, then you
will not know what the correct location should be, so you will not be able
to return the clip to the correct position using Spot mode.

When you click 'OK' in the Spot Dialog, the clip will be moved back to the
location where it was originally recorded – see Figure 6.20.

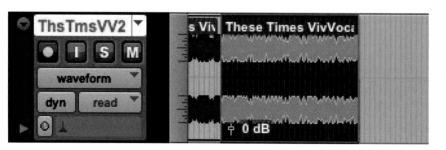

Figure 6.20
Edit window showing
Clip back in its
original position.

Grid Mode

Grid mode lets you constrain your edit selection to gridlines that correspond to a grid value that you can choose to suit your purpose. Grid mode is particularly useful if you are editing pattern-based music that starts and ends cleanly at regular boundaries, such as bars or beats. You can choose the Grid size using the Grid value pop-up menu located above the Timebase Rulers and the tracks in the Edit window.

> **TIP**
> If you press and hold the Control and Option keys (Mac) or the Start and Alt keys (in Windows), you can use the plus (+) and minus (–) keys on the numeric keypad to increment or decrement the grid size.

The Grid size can be based on a time value using the Main Time Scale; or, if Follow Main Time Scale is deselected in the Grid value pop-up menu, another time format can be used for the Grid size. To make the Grid lines visible in the Edit window, just click on the currently selected Timebase ruler name (the one highlighted in blue) to toggle these on and off.

> **TIP**
> You can temporarily suspend Grid mode and switch to Slip mode by holding down the Command key (the Control key in Windows), which is very useful while you are trimming clips, for example.

Absolute and Relative Grid Modes

Grid mode can be applied in either an Absolute or Relative way. In Absolute Grid mode, clips are 'snapped' onto Grid boundaries when you move them – so clips can never be placed in between the currently applicable Grid boundaries. This is what you normally expect and want Grid mode to do and is the default behaviour.

Relative Grid Mode lets you edit clips that are not aligned with Grid boundaries as though they were. For example, in 4/4 time signature, if a clip's start point falls between beats and the Grid is set to 1/4 notes, dragging the clip in Relative Grid mode will preserve the clip's position relative to the nearest beat.

In a music recording, for example, a musician may have deliberately played a note just before the beginning of a new bar. If you separate this note into its own clip, then use the Grabber tool to move it earlier or later by an exact number of beats using the Absolute Grid mode, the clip's start position will be 'snapped' exactly onto the new beat position and will not sound as the musician intended it to.

If you choose Relative Grid mode and set a suitable Grid value that allows sufficient time before the bar, then when you move the clip containing the note along the Grid, it 'snaps' to positions that preserve the note's original positioning just a little before the beat.

To select Relative Grid mode, click and hold the Grid button, move the mouse to where it says 'Relative Grid', and then let go – see Figure 6.21.

Figure 6.21
Selecting relative grid mode.

As you can see from the accompanying screenshot, see Figure 6.22, the bass plays a 'pickup' note before the beginning of bar 3. If you dragged this whole clip to bar 5 with Grid mode set to Bars and using Absolute Grid mode, it would 'snap' the beginning of the clip right onto the bar line. This would make no sense, because the 'pickup' note needs to play before the bar line.

Figure 6.22
The bass has a 'pickup' note.

The solution is to use Relative Grid mode instead. In this case, when you use the Grabber tool to slide the clip to right or left along the Timeline to a new position, the beginning of the clip will snap to a position that is exactly the same length before the bar line as the clip was positioned originally, keeping the pickup note's positioning relative to the bar line the same.

In the example provided, Figure 6.23, I dragged the clip later by two bars in Relative Grid Mode, and as you can see from the screenshot, it is now positioned a little before Bar 5.

Figure 6.23
Note dragged along the Timeline in Relative Grid mode to the correct position before Bar 5.

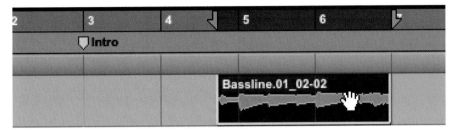

Dragging the clip in Absolute Grid mode would have only allowed me to position the start of the clip containing the note exactly onto a grid position – which is set to one bar. When I do this, as you will see from the accompanying screenshot, Figure 6.24, the note starts exactly on a bar grid line – which would, of course, sound wrong.

Figure 6.24
Note dragged along the Timeline in Absolute Grid mode snaps to a new (incorrect) bar position.

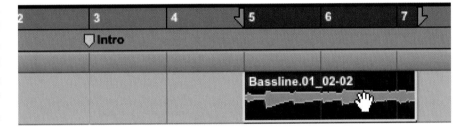

Editing in the Edit Window

The Pro Tools Edit window lets you edit audio non-destructively – editing a visual representation of the audio data without altering the audio source files. The clips that represent the audio can be cut, copied, pasted, trimmed, or cleared from the Edit window with no risk that you will permanently alter or lose any of your precious audio recordings.

There are some tools and processes that do work destructively – in other words, that can permanently change audio files on your hard disk. One important example is when you use the pencil tool to redraw an audio waveform, which does alter the audio file on disk when you save your changes. Another example is when you use the AudioSuite processes – but these are normally accompanied by clear warnings to this effect.

One of the great things about Pro Tools is that you can carry out many editing tasks while the session is playing back. You can separate and trim clips, place, spot or rearrange clips, add fades or crossfades, transpose or quantize MIDI tracks, nudge audio or MIDI clips, listen to different playlists, insert real-time plug-ins, and, perhaps most usefully, edit automation breakpoints. This makes it extremely fast to edit sessions – which can be a crucial factor on busy commercial recording sessions.

Whenever you record or import audio, video or MIDI data to tracks in Pro Tools, clips are created that represent this data visually in the tracks. Clips indicate where the material begins and ends and also give useful visual feedback about the general shape and content. New clips are created when you resize or separate, cut, or paste existing clips, and all clips of all types are displayed in the Clips List from which they can be dragged into tracks in the Edit window.

> **TIP**
> A clip can represent a complete audio file or just part of an audio file. For example, if you import a track from a CD, a clip is created that represents the whole track. If you then edit this, say, by trimming the clip in the Edit window, a second clip will be created that represents less than (by the amount you trimmed) the whole track.

Edit Window Track Views

By default, the Edit window is set to show tracks in 'Waveform' view. This means that you see a representation of the audio waveform presented as a graph of amplitude versus time. Audio tracks can also be set to show Playlists, Blocks, Volume, Volume Trim, Pan, Mute, or any plug-in controls that have been enabled for automation. You can also select the Elastic Audio Analysis and Warp views here. The default view for Audio tracks is Waveform view.

Auxiliary inputs can be set to Volume, Volume Trim, Pan, Mute, or plug-in automation controls; Master Fader Tracks can be set to Volume, Volume Trim, or plug-in automation controls; VCA Master Tracks can be set to Volume, Volume Trim, or Mute. The default view for VCA, Auxiliary, and Master tracks is Volume automation view.

MIDI tracks are usually set to either Notes or Clips and can also be set to Blocks, Volume, Pan, Mute, Velocity, Pitch Bend, After Touch, Program, Sysex, or any continuous controller type. Instrument tracks, like MIDI tracks, are usually set to either Notes or Clips and can also be set to Blocks, Volume, Pan, Mute, Velocity, Pitch Bend, After Touch, Program, Sysex, or any continuous controller type as well as Volume, Volume Trim, Pan, Mute, or any plug-in controls that have been enabled for automation.

The default view for MIDI and Instrument tracks is Clips view, which is the most useful view for editing and arranging MIDI clips in the Edit window.

> **TIP**
> Block view can be useful when you are cutting and pasting clips and want to move them around to rearrange them.

> **NOTE**
> For detailed MIDI editing, you can double-click any MIDI clip to open it in a MIDI Editor window.

Master Views

Audio tracks have two Master Views – Waveform and Blocks; MIDI and Instrument tracks have three – Clips, Blocks, and Notes (when using the Selector tool); and Auxiliary Input, Master Fader, and VCA Master tracks just have one – Volume.

When a track is in a Master View, any edits that you perform apply to all data in the track – including all the automation data. If you are working with data from multiple tracks and any of the selected tracks is set to a Master View, any edits you make will not only affect the audio or MIDI data on the track, but they will also apply to any automation or controller data on the track. So, when you are in a Master View, cutting an audio clip also cuts any volume, pan, mute, send, or plug-in automation that is also on the track – saving you from having to individually cut data from each automation playlist on the track. Similarly, when an Auxiliary Input or Master Fader track is displayed in its Master View, any edits performed apply to all automation data in the track.

In any other track view with automation data displayed on the selected tracks, edits only affect the type of automation data displayed in each track. For example, if track 1 displays Pan automation, track 2 displays Volume automation, and track 3 displays Mute automation, the Cut command cuts only pan data from track 1, volume data from track 2, and mute data from track 3.

NOTE
When Audio tracks are set to display either Waveform or Blocks, they are said to be in a Master View. When MIDI or Instrument tracks are set to display Clips, Blocks, or Notes (when using the Selector tool), these tracks are said to be in a Master View. When Auxiliary inputs are set to display Volume, this is regarded as the Master View, and Master Fader tracks (which can only display Volume) are always considered to be in Master View.

TIP
Pro Tools lets you override this behaviour temporarily by pressing and holding the Control (Macintosh) or the Start key (Windows) while you choose the Cut, Copy, or Paste commands, enabling you to copy all types of automation on all selected tracks.

Breakpoints

Breakpoints are points that can be inserted onto the graph lines that control automation data in the Pro Tools Edit window. As the manual explains, 'When an audio or Instrument track is displayed as Volume, Pan, or another automated control, or when a MIDI or Instrument track is set to one of the continuous controller types (Volume, Pitch Bend, After Touch), the data for that track appears in the form of a line graph with a series of editable breakpoints. The breakpoints can be dragged to modify the automation data, and new breakpoints can be inserted with the Pencil tool or a Grabber tool'. Got that?

All I would like to add to this is that editing volume and other automation data can often be done much more quickly and much more accurately by editing the breakpoints than by moving faders and other controls – which tends to generate lots of unnecessary automation data. Of course, if you like the feel of faders, knobs, and switches, nothing is going to substitute for these!

Switching Track Views

You can switch the view of any track using its Track View pop-up menu. When you have the Track Height set to Medium or greater, this is located to the left of the track in the Track Controls display area, just underneath the Record,

Solo, and Mute buttons. Otherwise, you can select this from the Track Options pop-up menu at the top left of the Controls area.

> **TIP**
> Use keyboard command shortcuts to get to the views you want. To change to the previous or next Track View, click in the track you want to change. If this track is a member of a group, all the other members will be selected. If you want to change views on more than one track, shift-click or drag the Selector tool to select additional tracks.
>
> With the track or tracks selected, just press and hold the Control and Command keys (Mac) or Control and Start keys (Windows); with these keys held down, you can use the left or right arrow keys on your computer keyboard to select the previous or next Track View.
>
> To change the view between Waveform and Volume (audio) or Notes and Clips (MIDI), just select the relevant track or tracks, as above and then press the Control and Minus (Mac) or Start and Minus (Windows) keys on your computer keyboard.

Useful Keyboard Commands for Editing

Pro Tools provides a useful keyboard command that lets you 'collapse' an Edit (or Timeline) selection and simultaneously move the Edit cursor insertion point to the beginning or end of the selection. Pro Tools also provides several keyboard shortcuts for moving and extending or decreasing the range of an Edit (or Timeline) selection. The best way to learn how these work is to make a selection, as I have in the accompanying screenshots, and try these for yourself.

Take a look at the following screenshot (Figure 6.25), which shows the Bar|Beats ruler above an audio track. Here you will see that I have selected part of a clip and you can see that the Timeline and Edit selection are linked, so the same four bars are selected in the Bar|Beats ruler. Notice that the Timeline Selection Start and End Markers are shown in red. Normally these will be blue, but when a track is record-enabled, these are shown in red to warn you. Notice also the small yellow flags positioned one bar before and one bar after the selection; these represent the Pre-roll and Post-roll amounts that are set in the Transport window and are yellow, as opposed to white, to signify that they are active.

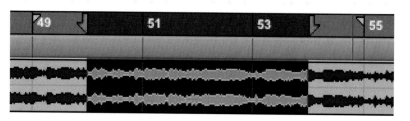

Figure 6.25
A clip selected in the Edit window with Timeline Selection Start and End.

Markers shown in red in the Bars|Beats ruler (because a track happens to be record-enabled).

While the Transport is stopped, you can use the Down and Up arrows on your computer keyboard to place the Edit cursor at the beginning or end of the Edit (or Timeline) selection and collapse the selection.

To move the insertion point to the beginning of the selection, press the Down arrow, or to move the insertion point to the end of the selection, press the Up arrow. You will immediately 'lose' the selection, which is then said to have 'collapsed' – see Figure 6.26 – and the insertion point will appear at the beginning or end of the selection.

Figure 6.26
Selection 'collapsed', with the Edit cursor inserted at the end of the selection and the Timeline Selection Marker placed above this in the Bars|Beats ruler.

To halve the length of the selection, press Command-Control-Option-Shift-L (Mac) or Control-Alt-Start-Shift-L (Windows) – see Figure 6.27.

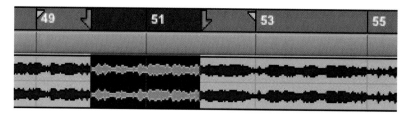

Figure 6.27
The Selection has been halved.

To move the selection forward by the selection amount, press Command-Control-Option-' (single quote) (Mac) or Control-Alt-Start-' (single quote) (Windows) – see Figure 6.28.

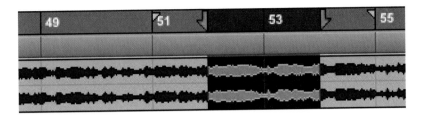

Figure 6.28
The Selection has been moved forward by the selection amount.

To move the selection backward by the selection amount, press Command-Control-Option-L (Mac) or Control-Alt-Start-L (Windows) – see Figure 6.29.

Figure 6.29
The Selection has been moved back again.

To double length of the Edit selection, press Command-Control-Option-Shift-' (single quote) (Mac) or Control-Alt-Start-Shift-' (single quote) (Windows) – see Figure 6.30.

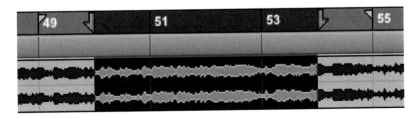

Figure 6.30
The Selection has been doubled.

To align the start point of a clip to timeline position, select the Grabber tool and use the mouse to point at and click in the timeline ruler to position the Edit cursor at exactly the point at which you wish to align the start point of the clip. Then, while holding down the Control key (Mac) or Start key (Windows), click on the clip and watch it snap exactly to the timeline position that you have specified!

NOTE

This trick even works when you are dragging clips into the timeline from the Clips List and can be faster to use than Spot mode (which lets you do much the same thing).

To align the end point of a clip to timeline position, select the Grabber tool and use the mouse to point at and click in the timeline ruler to position the Edit cursor at exactly the point at which you wish to align the end point of the clip. Then, while holding down the Command and Control keys (Mac) or Control and Start keys (Windows), click on the clip and watch it snap exactly to the timeline position that you have specified.

NOTE

This can be very useful if you want to align two sound files whose end point is the same but whose start points are different.

To align the sync point of a clip to timeline position, select the Grabber tool and use the mouse to point at and click in the timeline ruler to position the Edit cursor at exactly the point at which you wish to align the sync point of the clip. Then, while holding down the Control and Shift keys (Mac) or Start and Shift keys (Windows), click on the clip and watch it snap exactly to the timeline position that you have specified.

NOTE

This is particularly useful for aligning sound effects that have a sync point somewhere other than at the start of the clip, such as explosions or door slams.

Working with Clips

About Clips

Clips are the basic building blocks for arranging audio and MIDI in the Edit window, so you need to make sure that you know how these are created, edited, and arranged.

When you record new audio or import existing audio files, Pro Tools creates a Clip that plays back the entire file when placed into a playlist on a track. Very often you will use the Trimmer tool to remove some audio from the start or end of the recording. This creates a new Clip in the playlist that is shorter than the original clip, and this new clip also appears in the Clips List.

New clips are often created automatically when you make edits to existing clips. For example, if you use the Clear command to remove a section from within a clip, the sections on either side of this will form new clips. If you use the Cut command instead, the section you are cutting is placed onto the Clipboard and a new clip representing this section is also added to the Clips List.

Once you have created a clip, it appears in the Clips List. From the Clips List, you can drag it to a track to add to an existing arrangement of clips or to create a new arrangement 'from scratch'. You can slide clips or groups of clips around freely in the Edit window using the Time Grabber tool in Slip mode. You can also move clips around while constrained to the grid in Grid mode or shuffle them around in Shuffle mode or 'spot' clips to exact locations using Spot mode.

Capturing Clips

The 'Capture…' clip command defines a selection (see Figure 6.31) as a new clip and adds it to the Clips List. From there, the new clip can be dragged to any existing tracks.

Figure 6.31
A selection within a clip.

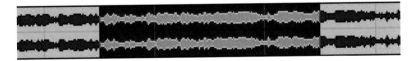

When you make a selection within a clip and choose 'Capture…' from the Clip menu or press Command-R (Mac) or Control-R (Windows), the Name dialog appears – see Figure 6.32. You can type a name for the captured clip here and OK this.

This captured clip then appears in the Clips List from where it can be dragged and dropped into any track in the Edit window at any location – see Figure 6.33.

Separating Clips

You can separate clips with the Selector tool enabled by inserting the cursor anywhere in a clip and pressing Command-E (Mac) or Control-E (Windows) or by

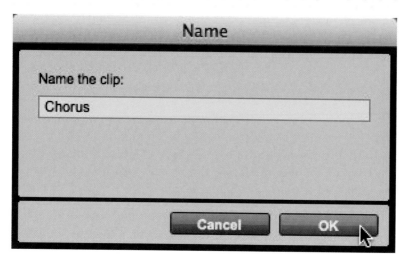

Figure 6.32
Naming a selection
within a clip using the
Clip menu's Capture…
command.

Figure 6.33
The captured Chorus
clip can be dragged
from the Clips List and
placed anywhere in
the Edit window –
here it is being
repeated.

choosing 'Separate Clip – at Selection' from the Edit menu. This simply cuts the clip into two at the insertion point. If you make an Edit selection within a clip and use this command to separate the clip, this will create two separate clips or three if you leave parts of the original clip unselected either side of your selection.

If you need to separate a clip into lots of smaller clips, you can use one or other of the two additional choices available in the Separate Clip sub-menu: 'Separate Clip On Grid' separates clips based on the currently displayed Grid values and boundaries. 'Separate Clip At Transients' separates clips at each detected transient.

Using either of these commands opens a 'Pre-Separate Amount' dialog that you can use to type a pre-separate amount in milliseconds to pad the beginnings of the new clips with a small extra amount of audio – see Figure 6.34.

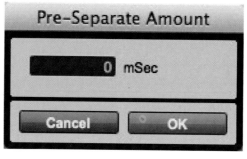

Figure 6.34
Pre-Separate
Amount dialog.

The Auto-Name Separated Clips option in the Editing Preferences page (see Figure 6.35) is selected by default, so Pro Tools automatically names separated clips for you.

If this preference is disabled, a dialog box appears that lets you type a name for the new clip.

Figure 6.35
Editing preferences
for clips.

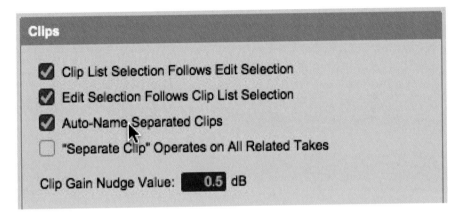

Figure 6.35
Editing preferences for clips.

NOTE
With looped clips, the Separate Clips commands automatically un-loop and flatten the looped clips before separating them.

Using the Separation Grabber

You can also create new clips using the Separation Grabber tool. Using the Separation Grabber saves you the trouble of separating the clip first. Make a selection using the Selector tool. Then select the Separation Grabber tool from the pop-up selector that appears when you press and hold the Grabber tool's button. Take a look at the accompanying screenshot (Figure 6.36) to see how this might look.

When you have selected the Separation Grabber tool, you will see a small pair of scissors appear on the 'hand' to remind you that the Separation Grabber is currently selected. Now you can drag your selection to a new location within

NOTE

When the 'Separate Clip' Operates on All Related Takes option is enabled in the Editing Preferences, editing a clip with the Separate Clip command also affects all other related takes with the same User Time Stamp. This option helps you compare different sections from a group of related takes. For example, you can quickly separate an entire group of related vocal takes into sections, then audition, and select the best material from each section independently. If this option is selected, make sure the Track Name and Clip Start and End options are also selected in the Matching Criteria window. If they are not, *all* clips in the session with the same User Time Stamp will be affected. In most instances, you will want to disable this option, to prevent a large number of clips from being created when you use the Separate Clip command.

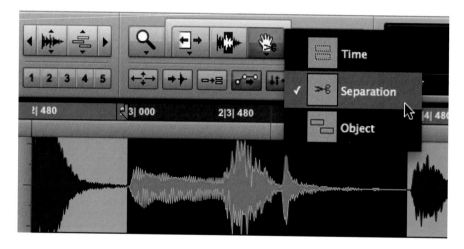

Figure 6.36
Choosing the Separation Grabber tool.

the same track or on another track. The selection is automatically separated from its 'parent' clip and a new clip containing this selection is created. As usual, new clips are also created from the material outside the original selection. Take a look at the accompanying screenshot (Figure 6.37) to see how this might look.

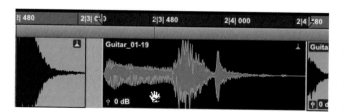

Figure 6.37
Using the Separation Grabber tool to separate a selection from a clip and move this new clip to a new location.

Healing Separated Clips

If you have made a cut in a clip or removed a section from within a clip, you may change your mind about this and want to return it to its original condition. If the clips have not been moved around since you made the cut or removed material, you can join the separated clips back together again using the Heal Separation command. The Heal Separation command returns separated clips to their original state – provided that the clips are still next to each other and that their relative start/end points haven't changed since they were separated.

> **TIP**
> If you want to leave the original clip intact, press the Option key (Mac) or the Alt key (Windows) while you drag the selection to the new location. A new clip containing a copy of the previous selection is created and placed at the new location while the original clip is not moved or changed in any way.

To heal a separation between two clips, use the Selector tool to make a selection that includes part of the first clip, the entire separation between the clips, and part of the second clip. Then choose Heal Separation from the Edit menu or press Command-H (Mac) or Control-H (Windows) to heal the separation.

> **NOTE**
> You cannot join two clips created from different audio files together using Heal Separation because they were never part of the same clip previously.

Overlapping Clips

Sometimes when you move clips around in Slip mode, you will inadvertently overlap these. You will also get overlapping clips on tick-based tracks with multiple clips if you increase the tempo. To warn you that this has happened, you can choose to display a small 'dog-ear' in the top right-hand corner of each clip if an overlap (or an 'underlap') has occurred.

TIP

Don't forget to select 'Overlap' in the Clip sub-menu from the View menu for this feature to operate!

In this example, I have copied and repeated a one-bar audio clip twice and then I dragged the second and third copies back along the timeline by one beat. The music is in 6/4 meter, so the first clip lasts 6 beats while the second and third clips appear to last 5 beats each, because they overlap.

The Clip menu also provides a couple of useful commands that let you bring an overlapped clip to the front or send it behind neighboring clips as necessary. If you select multiple clips and choose 'Bring to Front' from the Clip menu or press Option-Shift-F (Mac) or Alt-Shift-F (Windows), the first clip is placed in front of the second clip, the second clip is placed in front of the third clip, and so forth. Take a look at the accompanying screenshot (Figure 6.38) to see how this looks.

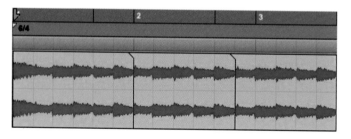

Figure 6.38
Three selected overlapping clips, each brought to the front before the next clip, so each overlaps the next.

If you select multiple clips and choose 'Send to Back' from the Clip menu or press Option-Shift-B (Mac) or Alt-Shift-B (Windows), the second clip overlaps the first clip, the third clip overlaps the second clip, and so forth. Again, take a look at the accompanying screenshot (Figure 6.39) to see how this looks. To describe this situation another way, you could say that each clip 'underlaps' the following clip.

Now if you select the middle of the three clips and send this to the back, the clip before and the clip after will both overlap this – see Figure 6.40.

And if you select the middle of the three clips, and bring this to the front, it will overlap both the clip before and the clip after – see Figure 6.41.

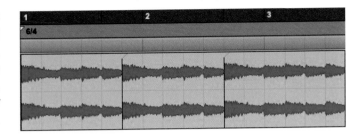

Figure 6.39
Three selected
overlapping clips,
each sent to the back
before the next clip,
so each 'underlaps'
the next.

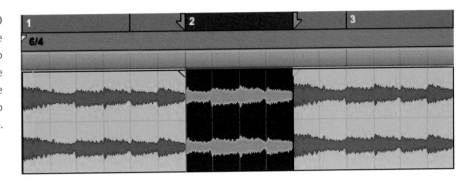

Figure 6.40
The middle of three
clips has been sent to
the back so that the
clip before and the
clip after both overlap
this clip.

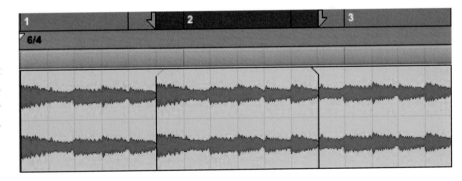

Figure 6.41
The middle of three
clips has been
brought to the front
so that it overlaps
both the clip before
and the clip after.

The easiest way to understand how all this works is to look at the accompanying screenshots, then try it for yourself.

Sync Points

When you are placing clips in Spot mode or Grid mode, it is sometimes useful to align a particular point within a clip with a specific Timeline location – instead of aligning the start point of the clip, which is the default situation. To cater for such situations, Pro Tools lets you define a sync point for any clip.

This situation often comes up when you are laying up music and sound effects to picture. The standard example quoted here is where you have a creaking door that eventually slams shut. The sound effect file includes the creak and the slam, and the obvious thing to do here is to line up the sound of the door slamming shut with the video frame at which the door actually slams shut – which is some way into the audio clip representing this sound effect.

TIP
To display sync points in your clips, make sure that Sync Point is selected in the Clip sub-menu that can be accessed from the View menu.

To define a clip's sync point, switch to Slip mode and place the Selector tool's insertion cursor at the exact point within the clip where you wish to position this. Then choose the Identify Sync Point command from the Clip menu to identify this as the sync point for the clip – see Figure 6.42. Alternatively, play the track and press the Down Arrow exactly where you want the sync point.

Figure 6.42
Setting a sync point within a clip with the Selector tool's insertion cursor placed at the position within the clip where the sync point is to be placed.

A small down arrow appears at the bottom of the clip, with a vertical, light grey line above this indicating the location of the sync point – see Figure 6.43

Figure 6.43
The sync point appears within the clip as a small green down arrow with a vertical, light grey line above this.

You can move this sync point to anywhere else inside the clip using the Time Grabber tool to drag it earlier or later. To remove a sync point, just Option-click (Mac) or Alt-click (Windows) on the sync point using the Time Grabber or Scrubber tools.

TIP

Hearing the audio scrubbing back and forth can often be far more revealing than looking at the waveform when you are trying to identify what sounds best. So you may find it helpful to use the Scrubber tool to help you locate where the sync point should be.

Nudging Clips

Pro Tools lets you nudge clips (or MIDI notes) by small increments or decrements along the tracks in the Edit window. The way this works is that you set a Nudge value using the Nudge Value pop-up menu (see Figure 6.44), then select a clip or group of clips; then you move (i.e. nudge) these forward or backward along the Timeline by pressing the plus (+) or minus (–) keys on the numeric keypad.

You can nudge material while Pro Tools is playing back, which really helps when you are fine-tuning 'grooves'. You can nudge continuously in real time to adjust the timing relationship between tracks and you can even nudge the positions of automation breakpoints in the playlists.

The Nudge value not only determines how far clips and selections are moved when you press the nudge keys, it can also be used to move the Start and end points for selections by the Nudge value or to trim clips by the Nudge value.

Figure 6.44 Setting the nudge value.

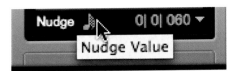

A pop-up selector near the top of the Edit window lets you choose the Nudge value – see Figure 6.45. With the main counter set to Bars:Beats, for example, the values offered are the common sub-divisions of a bar. You can also type the values you want directly into the Nudge Value display, which is useful if the values you want to use are not listed in the pop-up.

TIP

If you press and hold the Command and Option keys (Mac) or the Control and Alt keys (in Windows), you can use the plus (+) and minus (–) keys on the numeric keypad to increment or decrement the nudge value.

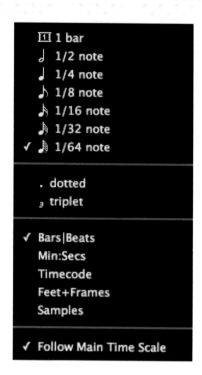

Figure 6.45

The nudge value

pop-up.

TIP

You can also nudge a clip's contents (sliding audio or MIDI into and out of the current clip boundaries) while keeping the clip's start and end points exactly where they are – assuming that there is material outside the clip's start and end points that can be slid into or out of the clip. Using the Time Grabber tool, select a clip whose contents you want to nudge. Press and hold the Control key (Mac) or the Start key (Windows) while you use the plus (+) and minus (–) keys to nudge the contents of the clip without changing the clip's start and end points.

Right-Click Commands for Clips

A really useful technique for editors is to use Right-click commands with key combinations to perform operations on objects while preserving selections in the Edit and Mix windows.

For example, after you have made a clip selection in the Timeline, or a Clip name selection in the Clips List, or a Track selection in the Edit or Mix window, you may wish to perform another operation without 'losing' this selection.

To preserve your selection while performing another operation in these circumstances, hold the Command key (Mac) or Control key (Windows) as you right-click the object and choose a command from the pop-up 'context' menu that appears – see Figure 6.46. (A different menu will appear depending on the 'context', i.e. on which object you are clicking.)

You can use this technique to open the Right-click Menu for one clip while you have another clip selected, for example.

Figure 6.46
As can be seen here, it
is possible to preserve
the selection of one
clip (at lower left)
while opening the
Right-click Menu of
another clip (at top
right).

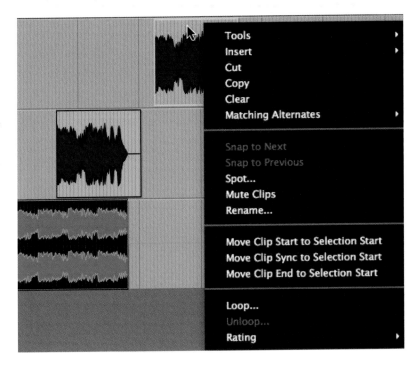

Or you can keep one track selected while carrying out an operation on another track. For example, if you hold the Command key (Mac) or Control key (Windows), you can open the Right-click Menu for one Track Name while a different Track Name is selected – as can be seen in the accompanying screenshot below, Figure 6.47.

Figure 6.47
While keeping the
Vocal track selected
(with its name
highlighted), it is
possible to open the
Track Name Right-
click Menu for a
different track such as
the Guitar.

Clip Editing Commands

The Edit menu and the Clip menu both contain useful commands that you can apply to clips.

Shift Clip

If you simply want to move selected material forward or backward in time along a track, and you know exactly how many bars and beats, seconds or samples, that you want to move this by, the quickest way is probably to use the Shift command. The Shift command works with selections, clips, MIDI notes, MIDI controller data, and automation breakpoints – and the selected material can reside on multiple tracks. To use this command, first select the track material you want to shift using the Selector or Time Grabber tool, then choose Shift from the Edit menu, or press Option-H (Mac) or Alt-H (Windows). The Shift dialog that opens, see Figure 6.48, lets you choose whether

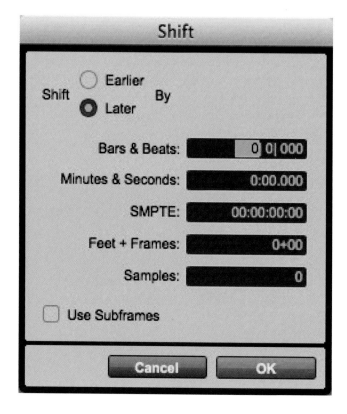

Figure 6.48
The Shift dialog.

the data will be moved earlier or later, and you can type the amount in bars and beats or using any of the other timebases.

> **NOTE**
> If you have selected just part of a clip, when you shift it, new clips are created from the selection and from any material outside of the selection.

Duplicate Clip

One of the most basic editing actions is to duplicate a clip. The Edit menu has a Duplicate command that copies a selection and places it immediately after the end of the selection. If you prefer to use the keyboard, press Command-D (Mac) or Control-D (Windows). You can achieve the same result using Copy and Paste, but using the Duplicate command is more convenient and quicker, especially if the clips are on more than one track.

> **NOTE**
> In Shuffle mode, the duplicate is placed directly after the end of the selection and any clips occurring after this will slide to accommodate the duplicated material. In Slip mode, the duplicated material will overlap any subsequent material.

If you are working in Bars|Beats, use Grid Edit mode and the duplicate will be placed immediately after the selection's end point – see Figure 6.49.

Figure 6.49
Using the Duplicate command: select a clip using the Selector tool, then choose Duplicate from the Edit menu, or use the keyboard command.

To copy a clip and automatically place it immediately before the selected clip, select the clip using the Grabber tool this time, then hold all three modifier

> **NOTE**
> When using Duplicate (or Repeat) for audio that must fall cleanly on the beat (such as rhythmic loops), it is important that you select the audio material with the Selector tool, or by typing in the start and end points in the Event Edit area. If you select an audio clip with the Time Grabber tool (or by double-clicking it with the Selector tool), the material may drift by several ticks because of sample-rounding.

keys, Command-Option-Control (on the Mac) or Start-Alt-Control (in Windows), and click once on the clip – see Figures 6.50 and 6.51.

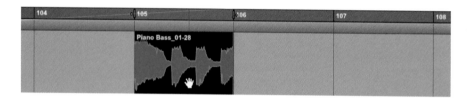

Figure 6.50
To duplicate a clip and place it before the selected clip, click using the three modifier keys and the Grabber tool.

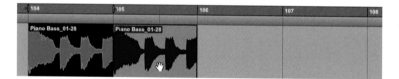

Figure 6.51
The duplicated clip is placed immediately before the clip that was clicked on.

And it gets better! If you click on another clip, located somewhere else in the track, so that this clip is selected first, then, using the Grabber tool, click on the clip you want to duplicate while holding the three modifier keys, a duplicate of this clip will attach itself immediately before the clip you first selected – see Figures 6.52 and 6.53.

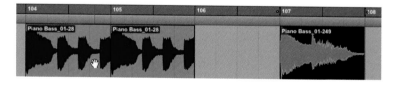

Figure 6.52
First select a clip before which you wish to place another clip so that it goes dark.

Figure 6.59
Nesting a clip group together with other clips into a new clip group.

When you choose the Group command, all three groups are joined together to form a new Clip Group – see Figure 6.60.

Figure 6.60
A new Clip Group created from an existing Clip Group together with two more clips.

The Clip menu also has commands to let you ungroup and regroup clips. If you apply the 'Ungroup' command to nested clip groups, it will only ungroup the top-layer clip group, leaving any underlying clip groups untouched. The 'Ungroup All' command – see Figure 6.61 – will ungroup a clip group together with all of its nested clip groups.

Figure 6.61
Using the Ungroup All command.

As you can see from the screenshot shown in Figure 6.62, after using the 'Ungroup All' command, the grouped clip is disassembled so that the original clips are accessible again.

Figure 6.62
Clips all ungrouped.

With the clips ungrouped you can edit these individually, as necessary. As long as you don't group and ungroup any other clips beforehand, you can then use the Regroup command to regroup these as they were and continue working on your arrangement.

You can also create multi-track clip groups by selecting clips on multiple adjacent tracks – see Figure 6.63.

Figure 6.63
Preparing to create
a mixed multitrack
Clip Group.

After selecting the clips, choose the Group command from the Clip menu
or press Command-Option-G (Mac) or Control-Option-G (Windows) – see
Figure 6.64.

Figure 6.64
A multi-track clip
group created by
grouping clips across
two tracks.

If you group a mix of sample- and tick-based audio or MIDI tracks together, as in
Figure 6.65, a different clip group icon in the bottom left corner of the clip group
indicates this.

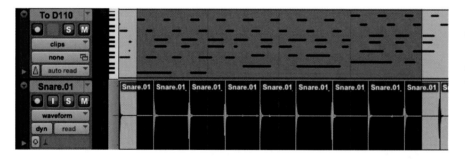

Figure 6.65
Selecting a sample-
based audio track
and a tick-based MIDI
track.

As you will see in the accompanying screenshot (Figure 6.66), when two
mixed clips are grouped like this, a special mixed Group icon is displayed in
the lower left-hand corner of the grouped clip.

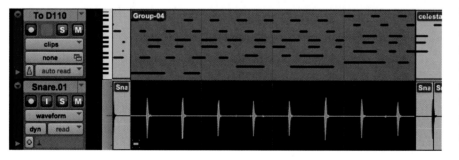

Figure 6.66
A mixed multi-track
Clip Group containing
both sample-based
and tick-based tracks
with a mixed clip
group icon in the
lower left corner.

NOTE

Clip groups can become separated if you move a track from the group so that it is no longer adjacent, hide a track from the group, insert a track within the group, delete a track from the group, record into a clip group, change the tempo of a mixed group, or change playlists on a track within the group. A broken icon appears in the lower left corner to warn you if this has happened.

In the example shown in Figure 6.67, the audio track is sample-based while the MIDI track is tick-based. When I changed the tempo after grouping these, the tick-based MIDI track moved along the timeline, while the sample-based audio clip stayed at the same location on its track. The 'broken' icon appeared in the lower left corner of both clips to warn that this had happened.

Figure 6.67
A mixed multi-track
Clip group separated
by changing tempo.

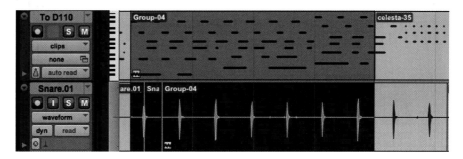

TIP

Pro Tools can export and import clip groups using the '.rgrp' clip group file format. This can be very useful for creating multitrack loops containing references to all the audio files within the clip group, clip names and relative location in tracks, track names, fades and crossfades, and all the MIDI data from the clip group. You can then import these into your current or future projects by dragging and dropping the clip group file from a Workspace browser or from Windows Explorer or Macintosh Finder to the Timeline, a track, the Track List, or the Clips List.

> **NOTE**
>
> Clip group files don't store automation, plug-ins, track routing, tempo or meter maps, or Clips List information.

Dynamic Transport Mode

Dynamic Transport mode lets you decouple the playback location from the Timeline selection so that you can start playback from anywhere on the Timeline without losing your Timeline or Edit selections. It is similar in this respect to using the 'Unlink Timeline from Edit Selection' feature, which is very useful for MIDI editing.

> **NOTE**
>
> When you enable Dynamic Transport mode, it disables the Link Timeline to Selection function. This allows you to play back either the Timeline selection or an Edit selection – whichever you prefer. You can press the left bracket key to audition the Edit selection and use the Enter key to play the timeline selection. This lets you carry on making your edits, very flexibly, while Pro Tools is playing back. And if you want the Timeline selection to follow the Edit selection, you can always re-enable the Link Timeline to Selection button whenever you like. Then the Dynamic Transport will follow each time you make a new selection in the Edit window.

You can enable Dynamic Transport mode by selecting this from the Options menu. If you prefer to use a keyboard command, press Command-Control-P (Mac) or Control-Start-P (Windows) to switch Dynamic Transport mode between its On and Off states.

When you are working in Dynamic Transport, there is a separate Play Start marker that you can move to a different location on the timeline – inside or outside of the timeline selection.

As soon as Dynamic Transport mode is enabled, the Main Timebase ruler expands to double-height and reveals this Play Start Marker, which marks the location from which playback will begin when you start playback.

The accompanying screenshot (Figure 6.68) shows the Play Start Marker located approximately halfway between bars 6 and 8 in this example, which

> **NOTE**
>
> The upper part of the main Timebase Ruler is referred to as the Timeline Selection Move Strip and the lower part is referred to as the Play Start Marker Strip.

has a time signature of 6/4. The Edit selection, as shown by the darkened clip in the Edit window and by the Edit Selection Start and End locations displayed in the Toolbar, does not play back when you press Play: the Timeline Selection, shown by the darkened area in the Main Timebase Ruler (the Bars|Beats ruler in this example) plays back instead.

Figure 6.68
When Dynamic Transport is enabled, the Main Timebase ruler expands to reveal the Play Start Marker.

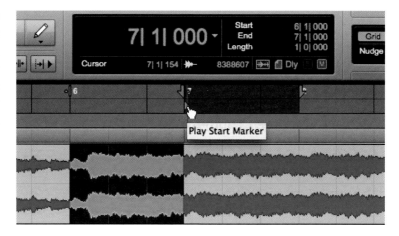

Here's an example of how you might use Dynamic Transport mode in Loop Playback mode:

If you move the Play Start to a location within the timeline selection that you are cycling around (see Figure 6.69), then start playback with Loop Playback enabled in the Options menu, it will start from this location, play to the end of the timeline selection, then subsequently loop around the timeline selection.

Figure 6.69
Play Start marker located within a Timeline selection.

NOTE

To reposition the Play Start Marker, you can point to any location within the Play Start Marker Strip in the Main Timebase ruler and click once, or you can simply click on the Play Start Marker and drag this to wherever you like along the timeline.

TIP

You can always position the Play Start Marker anywhere you like – it doesn't have to be at the start of the Timeline Selection or at the start of the Edit Selection.

A more practical scenario is to move the Play Start Marker to a bar or two before the timeline selection, so that it acts as a pre-roll to the cycle around the timeline selection.

In the accompanying screenshot (Figure 6.70), the Play Start Marker is positioned one beat before the Edit selection. When you press Play, playback will start from the Play Start Marker, play through the Edit Selection that follows this, play through the Timeline Selection that follows this, and then loop around the Timeline Selection until you stop playback.

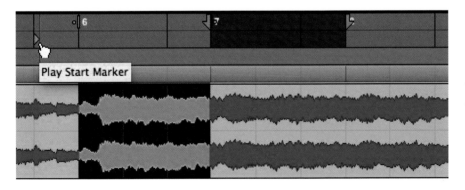

Figure 6.70
Play Start Marker positioned independently of the Edit Selection and of the Timeline Selection.

You can even reposition the Play Start Marker during playback and playback will continue from the new location.

So, if you want to judge how well the transition into the looped Timeline selection works, but you don't need to start from quite so far back next time, you can simply drag the Play Start Marker, during playback, to a new location – see Figure 6.71 – and when you let go of the mouse, playback will jump to that location and continue playback from there.

Figure 6.71
Dragging the Play
Start Marker to a
new location: this
even works during
playback.

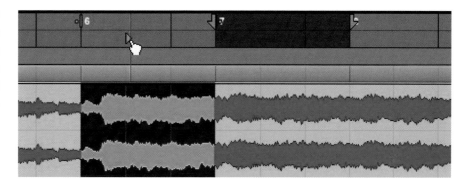

Dynamic Transport mode offers a lot more than simply unlinking the Timeline and Edit selections. For example, the Timeline selection can be freely moved around by pointing at the timeline selection with the mouse, then clicking and dragging the grey area that defines the selection in the Main Timebase ruler while holding down the mouse. Take a look at the two accompanying screenshots, Figures 6.72 and 6.73, to see how this looks: the mouse cursor changes to a 'hand grabber' icon when you point the mouse at the Timeline Selection Move Strip.

Figure 6.72
The Timeline selection
about to be dragged
from bar 6.

Figure 6.73
The Timeline selection
after being dragged
from bar 6 to bar 7.

Now that you know more about how Dynamic Transport mode works, here's an example of how you might use Dynamic Transport mode to cycle around a section of your song while you do some edits on-the-fly, then grab and

> **TIP**
> Remember also that you can always change the Timeline selection by clicking in the Main Timebase ruler using the Selector tool and dragging along the timeline to encompass a new selection. Alternatively, with any Edit Tool selected, you can just click and drag the Timeline Selection Start or End Markers to move these to new locations along the timeline.

drag the timeline selection to another part of the song to do more edits there:

Say you want to record a MIDI part in four-bar sections, each with a one- or two-bar pre-roll. To do this, you can set the Timeline selection to cover four bars, set the Play Start marker a bar or two before this, record this section, then grab and drag the Timeline selection to the next four bars, record this section, and so on. If you want to change the length of the Timeline selection, just drag the start or end markers for the selection back or forth along the Timeline so that it encompasses however many bars you want.

Keyboard Shortcuts for Dynamic Transport Mode

Because it is very likely that you will want to locate the Play Start Marker to the beginning or end of the Timeline or Edit selections, either during playback or when the Transport is stopped, there are several keyboard shortcuts provided for these and similar actions.

Pressing Period (.) on the numeric keypad then the Left Arrow moves the Play Start Marker to the Start of the Timeline Selection: pressing the Right Arrow moves it to the End of the Timeline Selection instead.

Pressing Period (.) on the numeric keypad then the Down Arrow moves the Play Start Marker to the Start of the Edit Selection.

You can nudge the Play Start Marker backward along the timeline, a bar at a time, by pressing 1 on the numeric keypad. To go forward, press 2.

Assuming that you have Bars|Beats selected as the Main Timebase ruler, you can move the Play Start to a specific bar using the numeric keypad: hold the Asterisk (*) key while you type the bar number, then let go of these keys and press the Enter key.

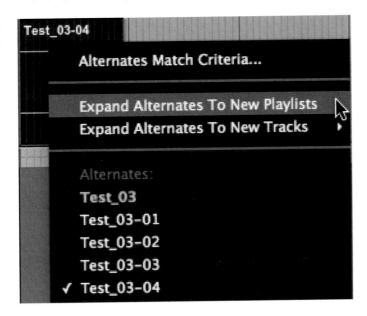

Figure 6.76
The Matches
sub-menu.

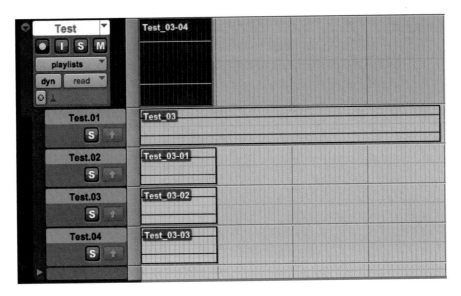

Figure 6.77
Four Alternate
Playlists revealed
below the Main
Playlist.

You don't really need to see the whole file clip that contains all the takes – you just want to see the individual takes that have the same start and end times as the Edit selection. To hide this long clip, Control-click (Mac) or Right-click (Windows) on the Playlist name and choose Hide – see Figure 6.78.

Figure 6.78
Hiding a 'take' in Playlists view.

Another way to hide a playlist lane is to select the same Edit selection that you used for loop recording from the main playlist, then Control-click (Mac) or Right-click (Windows) the Track Name or any of the Playlist Lane Names and choose 'Show Only Lanes With Clips Within the Edit Selection' from the Filter Lanes sub-menu – see Figure 6.79.

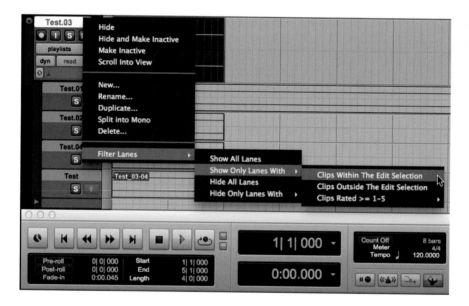

Figure 6.79
Filtering a Lane in Playlists view.

Choosing the Takes

When you have your takes showing in the alternate playlists in the 'Comp lanes' below the main playlist, you will probably want to audition these in Loop Playback mode, checking each take in turn by pressing the button marked 'S' on each Comp lane.

By the way, although the 'S' stands for 'Solo', pressing this does not solo the alternate take, it mutes the main playlist and plays the audio on the Comp lane instead – and you also hear the rest of your mix at the same time.

You can press Shift-S on your computer keyboard to solo any Comp lane containing the Edit cursor (see Figure 6.80), and you can use Control-P and Control-";" (Mac) or Start-P and Start-";" (Windows) to move the Edit cursor or the Edit selection up and down through the Comp lanes.

Figure 6.80
Soloing Comp Lane 'Test.03' to listen to a selection on this lane.

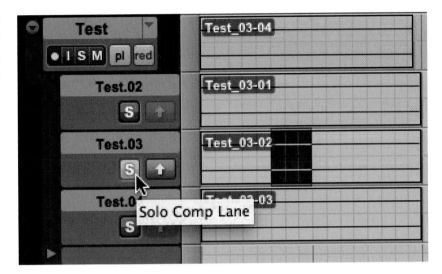

When you hear a section on a Comp Lane that you prefer, you can select this and click the upwards-pointing arrow to the right of the Solo button to copy this selection to the main playlist – see Figure 6.81.

Figure 6.81
Selection copied from playlist 'Test_03_02' to Main Playlist.

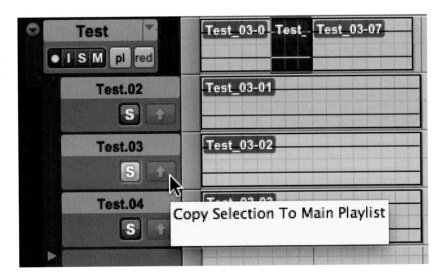

If you prefer to use a keyboard command, just press Control-Option-V (Mac) or Start-Alt-V (Windows) to copy the selection to the main playlist.

Working with Playlists

It is a feature that provides enormous flexibility in Pro Tools is its ability to create 'playlists'. Each track has a 'Main' playlist together with as many alternate edit playlists as you may wish to create. These Edit playlists contain whatever arrangement of clips you may have created in the track. You can easily duplicate the main playlist, then edit this to create alternate playlists, or simply create a new playlist and record new material into this playlist, and you can have a virtually unlimited number of playlists and share these among all tracks.

Each track has its own dedicated automation playlist. Automation playlists for audio, Auxiliary Input, Master Fader, and Instrument tracks store data for volume, pan, mute, and plug-in controls. Automation playlists for MIDI tracks, however, store only mute information; continuous controller events, program changes, and Sysex events are stored in MIDI clips and therefore reside within edit playlists for both MIDI and Instrument tracks.

When you create a new track, it contains a single, empty playlist until you record, import, or drag and drop material to it from the Clip List, Workspace browsers, Mac Finder, or Windows Explorer.

I find the playlists feature most useful when recording alternate takes of a new audio or MIDI tracks. Typically, I will record a guitar part into a track, then select 'New…' from the Playlist Selector – see Figure 6.82.

Figure 6.82
Creating a New
Playlist.

Playlists View

The default view for tracks shows the Main Playlist in Waveform view. Each track in the Edit window also has a Playlists view that shows the Main Playlist with any alternate playlists associated with the track displayed directly below the track in Playlist lanes.

These alternate playlists can be edited in the same ways that you can edit the main playlist in Waveform view. When you apply edits to range selections in Playlists view, these are applied to all alternate playlists that are shown. Alternate playlists that are not shown are not affected.

> **TIP**
> Before starting work with Track Compositing, duplicate the track's Main Playlist to keep it intact as a backup alternate playlist.

To view the Playlist lanes for a track, select Playlists from the Track View selector – see Figure 6.83.

Figure 6.83
Selecting Playlists
View from the Track
View Selector.

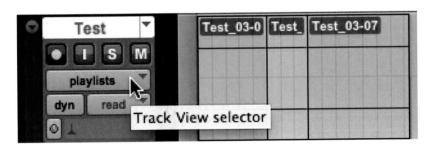

Alternatively, you can Command-Control-click (Mac) or Control-Alt-click (Windows) on the Playlist selector to switch to Playlists view. To switch back to Waveform view, hold these three modifier keys and click on the Playlist selector again – see Figure 6.84.

Figure 6.84
Click on the Playlist
Selector while holding
two modifier keys
to switch to Playlists
View.

If a track contains one or more alternate playlists in addition to the main playlist, these will be displayed in lanes underneath the main playlist.

If there are no alternate playlists, an empty playlist is available under the main playlist and can be used for adding clips (see Figure 6.85) from other audio tracks, or from the Workspace Browser or Clip List, to create new alternate playlists.

> **NOTE**
>
> If you have not created any alternate playlists for a track, you will still always see one empty playlist when you switch to Playlists view.

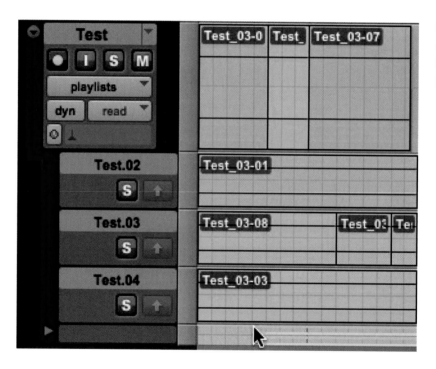

Figure 6.85
Adding another clip to the empty bottom playlist.

Copy Selection Commands

Once you have found a selection in an alternate playlist that you want to use in the main playlist, you can always manually copy and paste the selection to the main playlist to construct the best combination of selections from the different takes.

Pro Tools also provides three Edit menu commands that you can use to copy and paste the selection – see Figure 6.86.

Figure 6.86
'Copy Selection To....'
commands in the Edit
menu.

Figure 6.86
'Copy Selection To....'
commands in the Edit
menu.

Copy Selection to...	▶	Main Playlist	^⌥V
Strip Silence	⌘U	New Playlist	
TCE Edit to Timeline Selection	⌥⇧U	Duplicate Playlist	

The 'Copy Selection To Main Playlist' command copies and pastes the selection to the main playlist overwriting any material at that location. You can use this menu command as an alternative to using the buttons in each Comp lane.

The second command, 'Copy Selection to New Playlist', creates a new, empty main playlist and copies and pastes the selection to the new main playlist – see Figure 6.87. A dialog box opens when you use this command to let you type a name for the new playlist. I often use this command when I find the first section that I want to use, and name this as my 'Comp' track. Then I go through all the takes and just add the sections that I want to use to this playlist.

Figure 6.87
A new playlist, named
CompAudio, created
using the 'Copy
Selection to New
Playlist' command.

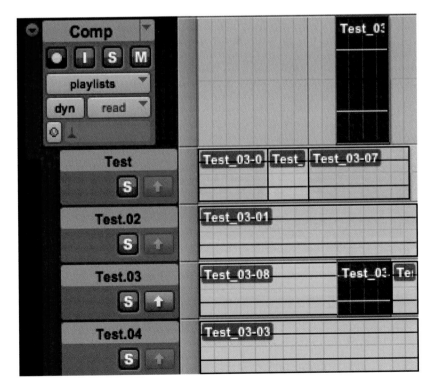

The 'Copy Selection to Duplicate Playlist' command works in a similar way, but in this case it duplicates the current main playlist and copies and pastes the selection to the duplicate main playlist. As with the 'Copy Selection to New Playlist' command, the previous main playlist moves to a new Playlist lane.

Why You May Need to Filter the Comp Lanes

Compositing tracks using these features can result in you having large numbers of playlists in the tracks you are working with. Say you loop record the vocalist on the verses and do four takes, then you do four more on the bridge sections, then four more on the chorus sections. Already that is 12 takes, and when you are comping the verse sections you probably don't want to view the bridges and choruses at the same time. This is where you will value the filtering feature.

You can filter Playlist lanes to show or hide them based on different criteria such as the Clip rating or whether they are inside or outside of the Edit selection. To filter Playlist lanes, you can either Control-click (Mac) or Right-click (Windows) the Track Name or the Playlist Name to access the context menu. If you access the menu from the Playlist Name, you get options to hide, scroll into view, create, delete or rename the playlist, or to Filter the Lanes – see Figure 6.88.

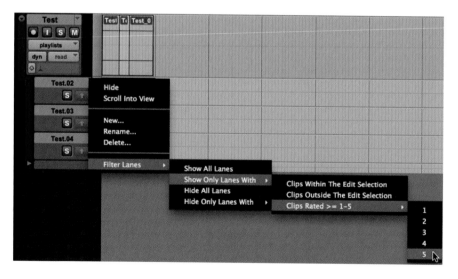

Figure 6.88
Control-clicking the Playlist name opens this menu.

You can choose a rating, from 1 to 5, for any clip by selecting the clip and choosing a rating from the Clip menu or by pressing Command-Control-Option (Mac) or Control-Alt-Start (Windows) and then typing a number between 1 and 5.

If you are going to filter based on ratings, it helps to view the ratings in the clips (as in Figure 6.89), and you can select this option from the Clip sub-menu in the View menu.

Figure 6.89

Viewing a rating in

a clip.

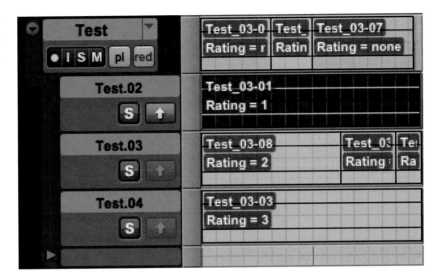

Compacting Audio Files

When you have finished your editing session, you should consider 'compacting' any edited audio files to remove any unused audio that you are sure you have no further use for. This makes the files smaller, so it saves disk space from being used up unnecessarily.

If you have recorded or imported a large number of audio files, yet have only actually used a fraction of these, compacting the files can save considerable amounts of disk space, making backups much less costly both in time needed to make the backups and for the media used.

To compact an audio file, select the clip or clips that you want to compact in the Clips List and choose Compact from the pop-up menu at the top right of the Clips List.

In the dialog window that appears, you can *pad* the clips within the compacted file to include some of the original audio (if this exists) at either side of the boundaries of the selected clips. You may have placed the clip boundaries too close to the edit points – cutting off very quiet sounds at the beginning or end of the clips. By including some of the audio just before and just after the clip boundaries, you ensure that this audio exists in the new file that you are creating – and you can always make more careful edits to remove this extra audio at any time in the future. So it is a wise thing to enter an amount of padding, in milliseconds, that you want to leave around each clip in the file.

When you click Compact, the file or files are compacted and the session is then automatically saved.

> **NOTE**
> Because it permanently deletes audio data, the Compact Selected command should only be used after you have completely finished your editing and are sure that you have no further use for the unused audio data.

Summary

The best way to learn how to edit in Pro Tools is, of course, to edit in Pro Tools. There is just no substitute for hands-on experience here. The best way to use this chapter is when you are sitting at your computer with Pro Tools right there in front of you so that you can try each editing move that I have described here for yourself.

Obviously, you are not going to learn everything in this chapter overnight, and you won't need to use all these editing techniques with every session that you work on. But if you go through this chapter thoroughly, trying as many of the editing commands and techniques as possible for yourself, there is a good chance that you will remember that a particular command or technique exists when the need for this arises.

You can also use this book as a handbook to keep around whenever you are using Pro Tools so that you can dip in and remind yourself of how a particular feature works when you need to use this.

In this chapter

CHAPTER 7

Mixing

Introduction

There are many approaches to mixing music that you will see recommended in books and by respected pundits. This chapter will not provide prescriptions for how to mix your music. Instead, the first section reveals various technical issues that you should consider while preparing for your mix. This is followed by discussion of the various 'tools of the trade' that you may use during your mixing sessions. The third section makes suggestions about how you might progress while you are organizing your mix, marking out the sections, ordering the tracks, grouping tracks together, setting up VCA groups, and using the Edit window to automate the mix. Finally, you get to the mix itself. You may be using a hardware controller together with the real-time Mix Automation features or you may mix 'manually' by setting the faders by hand, or by drawing in automation curves with the mouse. Ultimately, you will arrive at a result that you are happy with. At this point, you will make your 'final' or 'master' mix, which is the one that you will supply to the Mastering Studio that prepares the mixes for release on CD and in other formats. If you are intending to 'master' the mixes yourself, you may still choose to make these 'final' mixes at the end of the music production process and handle 'mastering' as a separate stage once production is finished.

Mixing is usually considered to be the final stage in the music production process, the end of the journey in many cases, but not always. In some genres of music, re-mixes are made for a variety of reasons. The most basic reason would be that the producer, artist, or record label feels that the first mix was simply not good enough. Another reason would be to provide the audiences with alternative versions to listen to when the music is broadcast on the radio or played by a club DJ. It is also quite common for another producer to hear the music, possibly years after it was first released commercially, and be enthused by the idea of creating a new mix – sometimes even a radically different mix that just uses the vocal and replaces all the music.

Preparing for Your Mix Session

Before you get started, you need to make sure that your monitors are set up correctly, consider what monitor levels to mix at, understand how your meters work, and line up the meters on any connected analogue equipment. To check mono compatibility, you need to understand how to use the Phase Scope, and if you want to avoid distortion problems that can arise with 'hot' mixes producing a 'fatiguing' sound that can be very unpleasant for the listener, then you need to know how an oversampling meter can be used to avoid such problems. If you are not using an external mixer, you will need a monitor controller to provide a convenient volume control for your playback listening level, and if you want to add an analogue 'sound signature' to your mix, one way to go is to use a summing mixer.

Monitoring

Check Your Monitoring

One of the most important things to attend to is your monitoring system – you have to be able to hear what you are doing properly when you are mixing. This means making sure that you are sitting in the 'sweet spot' at one of the three corners of an equilateral triangle with the left and right speakers positioned at the other corners facing you, and with the tweeters at about the same height as your ears.

You should make sure that the amplifiers have more than enough power to reproduce sound at the highest listening levels you intend to use without significant distortion and that the speakers have as 'flat' a response as possible over as wide a range as possible. So-called 'nearfield' monitors (such as the previously-ubiquitous Yamaha NS10's) that sit on stands, on desktops, or on the mixing console's meter bridge can never reproduce the full range of the audio spectrum – because they have to be relatively small. This means that they will not be able to reproduce the bottom octave properly as their response will roll off significantly below, say, 80 or 100 Hz. This is why professional studios have large, full-range monitors – to make sure that the engineers and producers can hear all the frequencies correctly.

You also need to take into account the acoustics of the room that you are listening in. For example, consider the situation where you are trying to set the level of a bass drum correctly for your mix. The fundamental frequency of this instrument might be around 80 Hz or even lower. The length of the sound wave that is produced at this frequency (the wavelength corresponding to this frequency) would be somewhere around 4 metres, which could easily correspond to the width of the room that you are listening in. A 'standing wave' would very likely exist at this frequency, sitting between the walls of your listening room. If you are sitting at a point where the amplitude of this wave is zero or very low, the bass drum will sound too quiet, and you will almost certainly compensate by making this too loud in your mix. This is exactly what happens in my home studio, for example. Conversely, if you are sitting at a position within the standing wave where the amplitude of the wave is at a maximum, you will think that the bass drum is too loud – and you are likely to compensate for this by making the level of the bass drum too quiet in your mix.

Rooms can be designed or modified to minimize these effects and monitoring systems can be adjusted to compensate for the room acoustics. Professional studios offer properly designed rooms and monitoring systems – for hire at professional prices. Smaller project studios and home studios often overlook, or the owners cannot afford, such necessities. If your listening environment and equipment is compromised, particularly at the low frequencies, then you will have to listen on other systems that will reveal problems at the low end (or elsewhere in the frequency ranges, for that matter).

TIP

I always check my mixes on the stereo system in my car, which instantly reveals whether I have made the bass drum and bass guitar too loud or not when compared with typical commercial albums. Then I go back and adjust the levels on my mixes, make another test CD and check this in the car, until everything sounds correct.

It is always a good idea to listen to your mixes on the kinds of speakers in the kinds of environments that will ultimately be used by your audience. So you might take your mixes out to a club, or play them in your car, or try them on your living room system – you get the idea. And don't forget to check your mixes in mono – especially if you want them to sound right on radio, TV, or film where mono playback is still sometimes encountered.

A Note on External Mixing/Monitoring

It is possible to connect a Pro Tools|HD audio interface directly to your monitoring system, but this is not a good way to set your system up because the only way to conveniently control the volume of your monitors would be to lower the Master Fader (or the channel faders).

If you lower the levels in your mix just to make the sound in the room quieter, then you are almost inevitably going to be making the level of your mix too low if you then save this as a file, or route the mix to an external recorder, or whatever.

Also, although there may be volume controls on self-powered monitors or on the power amplifier that is driving your passive monitors, these controls are not normally going to be easy for you to reach while operating the system.

As most engineers will be aware, Pro Tools, like most DAWs, is designed to give best results when the Master Fader that controls the summed output level is kept around unity gain (the 0 dB fader position) while the channel faders set the levels of the individual mix elements in relation to this. Although Pro Tools does allow you to lower the Master Fader substantially without significant loss of resolution, it is not good working practice to do so. For example, if the converters are operating at very low levels, any non-linearity will be most pronounced at these levels.

One solution is to use some kind of external monitoring control unit that allows you to connect the outputs from your Pro Tools interface and provides a volume control that controls its outputs to your monitors. Such units may allow you to connect two or more sets of monitors and switch between these, and may provide additional talkback, dimming, or foldback/headphone facilities.

Another solution is to use some form of external mixer that incorporates these facilities. Using an external mixer provides another advantage: it allows you to completely avoid any latency due to your Pro Tools system while recording from external sources, because you can route these into the external mixer, monitor them directly from the mixer (with no latency delay), and route them in turn into Pro Tools to record.

Most professional mixers also incorporate a number of reasonably high-quality microphone preamplifiers. These can be very useful if you are regularly recording from 'live' sources – and cost-effective compared with the price of buying lots of dedicated microphone preamplifiers.

Choose Your Monitoring Levels

There are no standard monitoring levels observed in music recording studios – unlike in film dubbing theatres, where the Society of Motion Picture and Television Engineers (SMPTE) has established a Sound Pressure Level (SPL) of 85 dB as a standard. This is normally referenced to an electrical signal level of −18 dBFS = 0 VU, which is the standard Operating Level recommended for digital systems by the Audio Engineering Society (AES). For the smaller rooms typically used for mixing music, a SPL of 83 dB may be more appropriate, with a nominal operating level of −20 dBFS providing 20 dB of headroom above this to accommodate signal peaks.

I have been at sessions in music studios that use monitoring levels well up into the 90s of decibels of SPL – or even more than 100 dB. Even relatively short exposure times to such high levels can make your ears 'sing' due to temporary threshold shift. Nevertheless, you may wish to check out the sound of some loud instruments that you have recorded at somewhere near to their original sound levels. And instruments such as the trumpet or snare drum can easily reach levels of over 100 dB, so it can be useful to have a monitoring system capable of reproducing such realistic levels.

Many engineers like to mix on small 'nearfield' monitors at quite low levels at somewhere between, say, 65 and 75 dB SPL. This makes it easier to tell whether the lead vocal or lead instrument can be heard properly at all times and whether other important elements such as the bass guitar and snare drum are at the right levels in relation to the lead. It also means that the mix should sound great when played back on domestic speakers after the record is released to the public. Of course, you can always check how it sounds from time to time at much louder volumes or on larger speakers.

> **TIP**
>
> I personally mix on 'nearfields' at around 83 dB or sometimes a little higher – checking this level occasionally using a small hand-held SPL meter. When I have a balance that is sounding good, I hit the dimmer switch on my Yamaha DM1000 mixing console to drop the monitoring level down by 20 dB to 60 or 65 dB, and then check that everything I want to hear is still audible.

Line-Up any Connected Analogue Equipment

If you connect your Pro Tools system to any analogue devices, such as mixers, tape recorders, or effects devices, you must make sure that the levels on the meters line up correctly. A 1 kHz test signal sent from Pro Tools at the chosen reference level, such as −20 dB relative to Full Scale, should produce a reading of 0 VU on an analogue VU meter connected to an analogue output from your Pro Tools interface. If it doesn't match exactly, you will need to 'tweak' the calibration control on the VU meter until it does match – this is the procedure for 'lining up' meters.

> **NOTE**
>
> Remember that the default meters in Pro Tools read signal peaks – they are not reading an average of the signal like a VU meter does – and there is no '0 VU' nominal operating level marked on peak meters.

> **TIP**
>
> Pro Tools HD 11 users can switch the metering to VU type if this is preferred.

To check that connected equipment is lined up correctly, you can insert the Signal Generator plug-in (see Figure 7.1) into one of the Master Fader Inserts, generate a 1 kHz sine wave test 'tone' at −20 dB, −18 dB, or whatever reference level is being used, and see whether the meters on connected equipment match this.

Figure 7.1
Using the Signal
Generator plug-in to
generate a test tone.

NOTE

If you connect your Pro Tools system digitally to external digital devices such as mixers, DAT (Digital Audio Tape) recorders, or effects units, then, if these use the same digital standards, signal levels sent from Pro Tools will automatically produce the same meter readings on these destination devices and vice versa. So, for example, audio at a level of −18 dBFS sent from a DAT recorder should show as −18 dBFS when it is received by your Pro Tools system and shown on its meters.

Figure 7.2
Recording a Test
Tone at −20 dBFS to
disk in mono from
a Signal Generator
plug-in inserted on
an Auxiliary Input to
an Audio track.

Recording a Test Tone to Disk

You can always record a test tone to disk, which can be useful if you will be transferring recordings to other systems, especially older analogue systems that need to be calibrated with their VU meters lined-up correctly.

To record a test tone to disk, simply insert the Signal Generator on an Auxiliary Input and bus its output to the input of an Audio track, as in Figure 7.2.

Creating a Test Tone Using AudioSuite

If you prefer, you can use the Signal Generator AudioSuite Plug-in to generate a 1-kHz sinewave at a suitable level to use as a reference tone so that other equipment can be aligned to in order to establish the correct value for 0 VU.

Make an empty Edit selection in an Audio track that lasts for 30 seconds, or however long you would like the test signal to last for, as in Figure 7.3.

Open the AudioSuite Signal Generator plug-in (see Figure 7.4) and choose the settings that you require.

Typically, you will choose a sine wave tone with an appropriate peak level such as −18 or −20 dBFS here, then press Render to create a file on disk containing the test signal, as in Figure 7.5.

You can supply this file along with your mixes – or record the audio from this in real time onto tape if you are transferring to tape.

> **TIP**
> The fastest way to create a test tone within a session after making an empty Edit selection in an audio track is to hold down the Control, Shift, and Option (Alt) keys then press 3. The Edit selection will immediately be filled by an audio Clip containing a 1-kHz sine wave at −20 dBFS – the default settings of the Signal Generator.

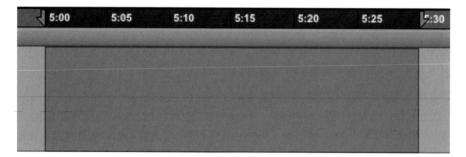

Figure 7.3
An empty 30-second timeline selection.

Figure 7.4
Using the Signal Generator AudioSuite Plug-in to generate a test tone.

Figure 7.5
Thirty seconds of
1 kHz sine wave
test tone.

Metering

VU Meters, Operating Levels, and Headroom

The 'VU' or 'Volume Units' meters used in analogue audio systems display an average measure of the signal level that corresponds reasonably well with the way the human ear perceives the volume of the reproduced audio. They do not reveal the peak levels of the audio signals, which may be 12 or more dBs higher than the average levels. Because these peak levels are typically very short in duration, they are often referred to as 'transients', meaning that they pass quickly.

The nominal Operating Level of an analogue recording or mixing system is the 0 VU level that is used as the standard input and output reference level for signals entering and leaving the systems. This 'nominal' operating level is the maximum average signal level that recording engineers using analogue equipment aim for when using VU meters, knowing that there will be 'headroom' above this to accommodate transient peak levels. The maximum peak levels that these systems can handle without distortion will vary, but would typically be at least 12 dB higher than 0 VU, as with tape recorders, for example. The difference between the 0 VU level and the maximum peak level is called the 'headroom'.

This operating level may be set differently in different audio systems and is typically set some way below the maximum level that the system can handle to allow 'headroom' for transient level peaks. If signals passing through the system are much lower than 0 VU, they may get too close to the background noise floor that exists in all audio systems. On the other hand, if average signal levels are much higher than 0 VU, transient peaks within these may attempt to exceed the maximum allowable level in the system, resulting in the waveforms becoming 'clipped' and the sound becoming distorted. In practice, you should set levels so that the loudest sounds average out around 0 VU without getting too much lower or too much higher. Also, don't forget that most music will normally contain many quieter sounds that may be 20 dB less (or even lower) in level than the loudest sounds.

Of course, you are not seeing the peak levels on VU meters, especially transient peaks that only occur for a fraction of a second, and peak levels can be much, much higher than average levels. To completely avoid distortion when recording audio to analogue tape systems, peak levels should not exceed 12 dB above 0 VU. A characteristic of analogue systems such as tape recorders is that the distortion is mild at first – so relatively high levels can be used before obnoxious distortion is heard. Some recording engineers exploit this situation by recording higher levels onto tape, so that some distortion of the waveforms occurs – making the recordings sound 'bigger' or 'fatter' than perfectly clean, un-distorted recordings.

Audio that is being recorded to a digital system from an analogue system with 0 VU set to correspond to −20 dBFS has 20 dB of headroom to accommodate signal peaks. However, in digital systems, if any signals exceed the 0 dB Full Scale level they will be 'clipped' to this level as they pass through the system – badly distorting the signal and making it sound extremely unpleasant. So with digital systems, you should always make sure that you have levels set low enough to avoid clipping – even if you have to use compressors or limiters to allow you to increase average signal levels.

> **NOTE**
> Pro Tools 11 will allow you to work internally, digitally, with audio signals at higher levels than 0 dB FS because of its 64-bit 'audio engine' with its 32-bit floating-point file structures.

When you are working with digital systems, the nominal Operating Level for any connected analogue audio equipment is sometimes set at 18 dB below the 0 dB Full Scale (0 dBFS) level that represents the maximum allowable signal level in the digital system. This provides headroom of 18 dB above the 0 VU level on the analogue equipment. Another standard that is recommended when you are mixing audio is to set the nominal Operating Level at −20 dB FS. This allows 20 dB of headroom for peak levels, encouraging the mix engineer to create mixes with greater crest factors (peak-to-average ratios).

Metering in Pro Tools

Each Pro Tools track has its own metering running vertically to the right of the track's fader. These meters let you see the signal levels on individual faders, or what the summed signal levels are at the Master faders. Audio tracks default to Pre-Fader Metering, so the meters on Audio tracks show the levels of signals being presented at the track inputs from the audio interface or being played back from audio files on disk. On Auxiliary Inputs, Instrument tracks, and Master Faders, the meters indicate the level of the signal being played through the channel output or outputs.

Meter and Fader Scales

The scale markings to the left of each *meter* when using Sample Peak Metering are peak levels that show 0 dB Full Scale at the top of the range. These markings will vary according to the meter scale type in use. For example, the meter scale for the Sample Peak meter type, see Figure 7.9, ranges from 0 dB down to −60 dB, while the meter scale for the VU meter type, see Figure 7.10, ranges from +3 dB down to −20 dB.

The scale markings to the left of each *fader* show a unity gain position (no increase or decrease in level) marked '0' (Zero). The fader controls the gain for the channel, and this can be boosted by up to 12 dB or attenuated by more than 90 dB, down to a minimum of –INF dB.

> **NOTE**
>
> A fader controls the gain, or amplification, applied to an audio signal. Increasing the gain above the unity gain level boosts, or amplifies, the signal. Decreasing the gain below the unity gain level attenuates the signal, reducing its level. The amount of positive or negative gain applied to a signal is simply expressed as a ratio in decibels of the output level compared with the input level.

If you set the fader at unity gain on an Audio track (zero on the scale), the playback level of the audio coming from the disk is not altered. You can increase this playback level by up to 12 dB or you can decrease it infinitely until you hear no audio playing through the channel – although you will still see the level being presented to the Audio track's input from the audio file on disk showing up, unaltered, in the Track meter.

> **NOTE**
>
> You can change the metering for all the Audio tracks in your session to post-fader metering by de-selecting the option for Pre-Fader Metering in the Options menu. In this case, the meters show the output level from the track, as controlled by the fader. So if you pull the fader down, the level displayed on the meter will go down accordingly. This is a less useful option, because it does not reveal any clipping that might be present at the input to the track.

If you have audio generated by a plug-in inserted into an Instrument or Auxiliary track or coming into an Auxiliary channel from an external source, the fader controls the gain for this in the same way. The meter shows the level being presented to the audio input of the Instrument or Auxiliary track from the external source or from the plug-in, and the fader controls how much of this signal is passed through to the track output (which usually feeds your mix bus, but could be routed wherever else you choose). So if you bring down the level of the track fader, the signal level in the Track meter does not alter. But if this track is routed to a Master Fader, for example, you would see the level in its meter fall in as you reduce the gain on either the track fader or the Master Fader.

> **NOTE**
>
> Nothing that you do with an Audio fader (that only ever controls the level of signals passed to the output of the track) will affect the audio that is in a file on disk playing back via this track. So if the audio that was recorded into a file on disk is clipped, for example, reducing the playback level will not prevent it from clipping – it will just play back the clipped audio at a lower level – and if the level of audio in the file on disk is sufficiently high, you will continue to see the clip indicator light up in the meter on the audio track.

Figure 7.6
Master Fader set to Sample Peak Metering.

Advanced Meters

One of the most welcome improvements for Pro Tools 11 is the new meter designs for audio, Auxiliary Input, Instrument, and Master Fader tracks in the Mix, Edit, Output, and Send windows. The meters are also longer than previously and provide more information.

Pro Tools 11 offers four basic types of meter: Sample Peak (see Figure 7.6 – the default meter type), Pro Tools Classic, Venue Peak, and Venue RMS.

With Pro Tools HD 11, you also get Linear and Linear Extended, RMS, VU, and Digital VU, three Bob Katz K-scale meters, several PPM Meter Types, and PPM Digital.

> **NOTE**
>
> Pro Tools 11 provides a new Metering page in the Preferences window. The Metering page includes preferences that were previously located in the Display preferences together with new settings for the advanced Meter Types and gain reduction meters.

Changes that Affect the Clip Indicators in Pro Tools 11

You may get a surprise when you look at the Clip indicators on tracks and sends in Pro Tools 11 because these display in Yellow – not Red!

Master Fader clip indicators (and those in the Output meters in the Transport) do still display in Red when the audio signal exceeds 0 dBFS to warn you that this is causing clipping at the output converters on your audio interface.

The reason that Avid decided that Audio track clip indicators will normally display yellow when the signal exceeds 0 dBFS is because the 64-bit floating-point calculations used will allow for such a wide range of values that the full range of possible audio levels can easily be accommodated – with no possibility of internal clipping!

However, it is still possible to clip at the converters on your audio interface or to clip when writing audio to disk in fixed-point 16-bit or 24-bit audio files (although 32-bit floating-point files will not clip).

So Auxiliary Input track, Instrument track, and Send clip indicators also display yellow when the signal exceeds 0 dBFS, even though, as we now know, there can be no internal clipping. This provides a warning that these audio streams can clip at the output converters or when writing fixed-point file formats to disk.

Also, when an audio track is record-enabled or set to input monitoring and the input assignment is set to a physical input signal that exceeds 0 dBFS, the clip indicators will still display in red to warn that this is causing clipping at the converters. The same thing happens when an audio track is recording to a fixed-point bit-depth (16- or 24-bit) and the audio written to disk exceeds 0 dBFS causing clipping in the file.

> **TIP**
>
> Best practice for mix engineers is to keep the highest peak levels below −3dBFS, even when bouncing to 32-bit floating-point files. Although the audio in a 32-bit file will not be clipped even if levels exceed 0 dBFS, it is possible that such levels could cause problems at later stages in the production process, especially if different engineers work on these files, and particularly if these are less-experienced engineers, possibly using less capable equipment.

Figure 7.7
'Fat' Meters.

Fat Meters

To see even more information, you can use the 'fat' meters instead of the normal 'thin' meters in the Mix and Edit windows, as in Figure 7.7. To switch between normal and 'fat' meters, hold the Command, Option and Control keys (Mac) or the Start, Option and Control keys (Windows) and click on any of the meters.

Gain Reduction Meter

With Pro Tools HD 11 software, a Gain Reduction (Dynamics) meter can optionally be displayed just to the right of the track signal meters on audio, Auxiliary Input, Instrument, and Master Fader tracks that are using supported dynamics plug-ins (such as the Avid Pro Compressor).

The Gain Reduction meter can be displayed whenever dynamics processing plug-ins are in use on the track. Because both compressor/limiter and expander/ gate plug-ins can be used at the same time on tracks, there are options in the Metering Preferences (see Figure 7.8) to allow the Dynamics meter to respond

Figure 7.8
Gain Reduction
Metering preferences.

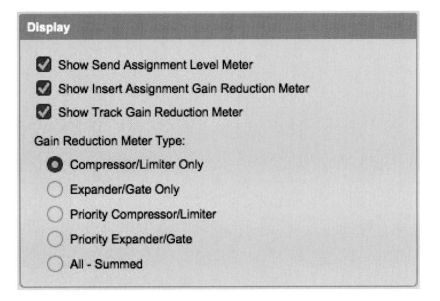

to either of these – or both. This Gain Reduction meter displays in orange to distinguish it from the Track signal meters which display green as can be seen in Figures 7.9 and 7.10.

Figure 7.9
Advanced Meters set
to the default Sample
Peak type.

Meter Type
scale

Fader Gain
scale

Gain Reduction
meter
(Pro Tools
HD only)

Track signal
meters

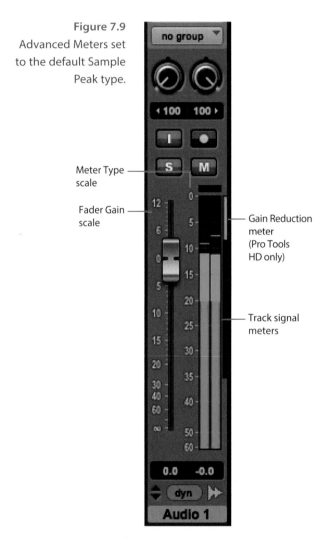

Figure 7.10
Advanced Meters set
to VU standard on
an Instrument track
showing the stereo
level meters and the
Gain Reduction meter
with its orange level
indicator.

Linking the Track and Master Meter Types

The Track and Master Meters Types section of the Metering Preferences window has a checkbox that lets you link the Track and Master Meter Types – see Figure 7.11.

Figure 7.11
The Track and Master
Meter Types section
of the Preferences
window – Linked
enabled.

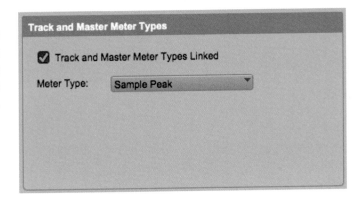

You can disable this option if you want the Track and Master meters to use different Meter Type settings (e.g. if you want to use Peak metering for your Track Meters and VU metering for your Master meters) – see Figure 7.12.

Figure 7.12
The Track and Master
Meter Types unlinked.

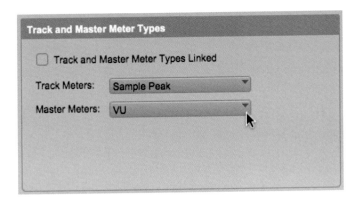

The Track Meters selector lets you choose the Meter Type for audio, Auxiliary Input, and Instrument tracks; the Master Meters selector lets you choose the Meter Type independently for Master Fader tracks.

Advanced Meter Type Settings

The Advanced Meter Type Settings section in the Metering Preferences window lets you adjust the scale and ballistics of the selected Meter Type – see Figure 7.13. These settings update automatically to match the specifications of the selected Meter Type. In some system configurations, you may want to make manual adjustments to these settings in order to ensure that Pro Tools Metering matches the meter response your broadcast console (some broadcast consoles use variations from the standard Meter

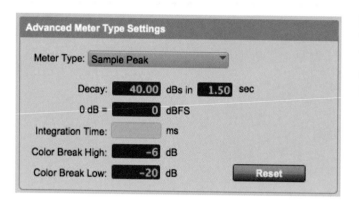

Figure 7.13
The Advanced Meter Type Settings section of the Metering Preferences window.

Types that are provided in Pro Tools). However, in most cases, these settings should be left at the default values.

If you are working with an external mixing console that uses variations from the standard Meter Types provided with Pro Tools, you can make manual adjustments to various settings to match these, as follows:

▶ Use the Decay settings to determine the decibel range (dBs) that the meters will fall at the specified rate in seconds after peak level is registered. For some Meter Types, these setting are not adjustable (such as VU).

▶ Use the 0 dBFS setting to determine the unity reference level for the meter scale in decibels compared to digital 0 dB. Exceeding this level should result in clipping at the digital-to-analogue converters or to the specified clipping level for broadcast. For some Meter Types, this setting is not adjustable (such as K-meters).

▶ Use the Integration Time setting to determine the amount of time it takes, in milliseconds, to average the peak level as registered on the meter display. You can think of this as the attack time for registering peak levels on the meters. For some Meter Types, this setting is not adjustable (such as Peak).

▶ Use the Crossover Hi setting to determine the high point in dB where the colour changes for metering. Typically, this will be the ceiling level for the current Meter Type. When this level is exceeded clipping will occur. Levels registering on the meters between the Crossover Hi setting and the Crossover Low setting should be considered the 'sweet spot' for the loudness of the program material.

▶ Use the Crossover Low setting to determine the low point in dB where the colour changes for metering. Typically, this will be the minimum level for the current Meter Type. Use this setting to provide a visual indication in the meters for the lowest level of the dynamic range of the program material for the intended delivery format (such as film, DVD, CD, MP3, or broadcast).

Meter Types

Using the Meter Type pop-up selector in the Advanced Meter Type Settings section of the Metering Preferences window, you can change metering ballistics and scales so that the Pro Tools Metering matches the metering on various external consoles – see Figure 7.14.

Figure 7.14
Pro Tools Metering
Preferences showing
the Meter Type
pop-up selector.

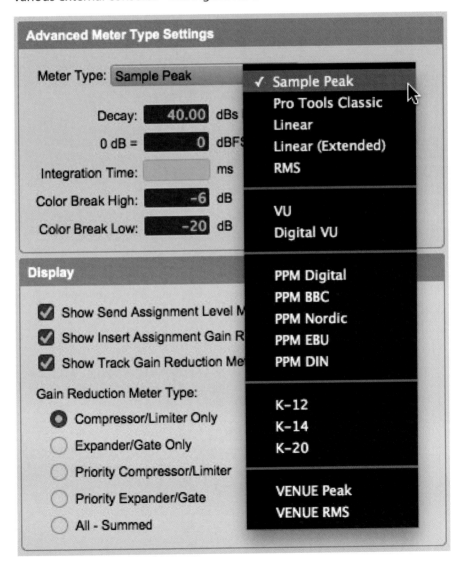

Options include a variety of PPM standards, VU metering, Bob Katz's K-scales, Sample Peak, and RMS (Average Loudness) metering.

Pro Tools Meter Types

Sample Peak is the default type of Pro Tools Metering that simply measures sample peaks – responding to every transient. This is the only Meter Type used for non-HD versions of Pro Tools. The scale and the decay time are calculated in dB/second, which results in slower decay in metering compared to lower versions of Pro Tools (such as Pro Tools 10). The Sample Peak option is the only Meter Type that has a 0 sample integration time, and as such shows all dynamic activity of the digital signal at every moment in time.

There are three other Sample Peak Meter Types available: Pro Tools Classic provides legacy Pro Tools scale and metering ballistics; VENUE Peak provides the same ballistics as Sample Peak, but with VENUE meter scaling to +20 dB; VENUE RMS provides the same ballistics as RMS, but with VENUE meter scaling to +20 dB.

> **NOTE**
>
> You can set the Meter Type either in the Pro Tools Metering preferences or by Right-clicking any Meter and choosing from the pop-up menu. If the Track and Master Meter Types are not linked, you can set the Meter Types for these independently.

> **TIP**
>
> To set all track types to the same setting, regardless of whether the Track and Master Meter Types are linked, Option-Right-click (Mac) or Alt-Right-Click (Windows) any track meter and select the Meter Type.

Pro Tools HD Meter Types

Linear is designed for use with post-production and music mixing scenarios. Using a fast decay time, Linear provides direct one-to-one linear metering of sample peaks in the audio signal with a metering range down to −40 dB.

This offers higher metering resolution closer to 0 dB (which can be particularly useful for mixing and mastering).

Linear (Extended) provides the same ballistics as Linear, but the meter scaling extends to −60 dB.

RMS provides metering ballistics that display the average loudness (root mean square of the signal) over a range of time. Sometimes this can be a useful alternative to Peak metering, which displays the peak signal level at any given point in time.

VU metering is popular for music and dialogue because it averages out the short-duration peaks and troughs and provides a better representation of the way the human ear perceives loudness. VU meters are specifically designed to have a dynamic characteristic that approximates the response of the human ear.

Digital VU provides VU ballistics with a modern digital scale.

PPM (Peak Programme) Meters, sometimes referred to as 'quasi-peak' meters, only show transients that are sustained for more than a few milliseconds. Engineers using these for music with analogue mixing consoles will often drive programme levels higher, consequently improving signal-to-noise ratios, while ignoring the inevitability that shorter transients will be clipped (they assume that this clipping will be inaudible).

PPM Digital is mostly used in Europe and Asia by broadcasters, and also with US Film consoles (such as the Avid System 5 consoles). PPM Digital has a similar integration time to Sample Peak Metering, but different scales and decay times. The PPM Digital option (like all other Meter Type options apart from Sample Peak) does not report every transient (such as a series of very fast transients). However, even though it under-reports the actual digital dynamic range, it more closely matches human hearing of loudness over time.

PPM BBC is used in the UK by commercial broadcasters. BBC scaling uses 4 dB spacing between scale marks. Other organizations around the world, including the EBU, CBC, and ABC, used the same dynamics but with slightly different scales.

PPM Nordic is a Scandinavian variant of the DIN PPM. This has the same integration and return times but a different scale, with 'TEST' corresponding to Alignment Level (0 dBu) and +9 corresponding to Permitted Maximum Level (+9 dBu). Compared to the DIN scale, the Nordic scale is more logarithmic and covers a somewhat smaller dynamic range.

PPM EBU is a variant of the British PPM designed for the control of program levels in international program exchange (Type IIb PPM in IEC 60268-10). It is identical to the British PPM (BBC) except for the meter scale. The meter scale is calibrated in dB relative to the Alignment Level, which is marked 'TEST'. There are ticks at 2 dB intervals and at +9 dB, which corresponds to the Permitted Maximum Level.

PPM DIN is the German broadcast standard. The nominal analogue signal corresponding to Permitted Maximum Level was standardized by ARD at 1.55 volts (+6 dBu), and this is the usual sensitivity of a DIN-type PPM for an indication of 0 dB. The Alignment Level (−3 dBu) is shown on the meter by a scale mark at −9.

Bob Katz's K-scales are RMS-based scales with an integrated sample peak meter as a secondary value. K-Scales are popular with music mixers that are looking for a meaningful indication of overall loudness. K-12 is for broadcast audio, K-14 is for mastering in a calibrated mastering suite, and K-20 is best used for mixing music.

Output Meters

To make it even easier for you to keep a watchful eye on the output levels at all times, Pro Tools 11 provides optional Output meters in the Transport – see Figure 7.15.

Figure 7.15
Transport Window showing Output Meters at the far right.

These Output Meters display the monitoring levels for signals routed to the physical outputs on your audio interfaces from your session. The top LED on each channel is red and this lights whenever clipping occurs at the digital-to-analogue converters on your audio interface.

You can show (or hide) the Output Meters in the Edit window transport or in the Transport window, by selecting (or de-selecting) Output Meter from the Edit window or Transport window menu.

Figure 7.16
Headroom indicator
with the arrow cursor
pointing towards this.

Headroom Indicator

Yet another convenient indicator has been added to tracks in the Mix window in Pro Tools 11 to give you feedback about the amount of available headroom. This Headroom Indicator provides a numeric display (in dB) of the available headroom on tracks based on the last peak level in relation to the reference level of the currently selected Meter Type – see Figure 7.16.

The Headroom indicator is located just to the right of the Audio Volume indicator – in the area immediately below the faders and meters in the Mix window. It holds the value of the highest peak level until a greater peak level arrives. To see the current peak level, just click the Headroom indicator to reset this to display the current value.

Expanded Sends and Mini Meters in Insert/Send Assignments

With all versions of Pro Tools 11, you can have an 'expanded' send level display in the send assignments view for each individual send. With previous versions of Pro Tools, only one send from any group of five could be displayed in this expanded view. With Pro Tools 11 you can expand any or all of the sends.

Figure 7.17
Here you can see
Mini Meters in the
Send and Insert
assignments. You can
also see the Expanded
Send view for Send A.

The expanded view provides more visual feedback (via the medium-sized send level meters) and is quicker to tweak, using its mini-fader and rotary pan controls. You can Command-click (Mac) or Control-click (Windows) on any send selector to 'expand' or 'collapse' the send. Similarly, you can Command-Option-click (Mac) or Control-Alt-click (Windows) on any send selector to expand or collapse all sends within the send group (i.e. Sends A–E or Sends F–J).

You can also enable the 'Show Mini Meters in Send Assignments and/or Insert Assignments' options to show small 'Mini' level meters in the Sends and/or Inserts Assignments on audio, Auxiliary Input, Instrument, and Master Fader tracks in the Mix window. These Mini Meters give you convenient visual feedback to confirm that audio is passing through the Send or Insert – see Figure 7.17.

NOTE

Mini Meters will only appear on tracks with supported dynamics plug-ins inserted.

One really useful view with big mixes is an 'all sends expanded' view or 'sends matrix'. Using the View menu, you could also hide the main channel meters and faders, which would allow you to focus your attention mostly on the sends – see Figure 7.18.

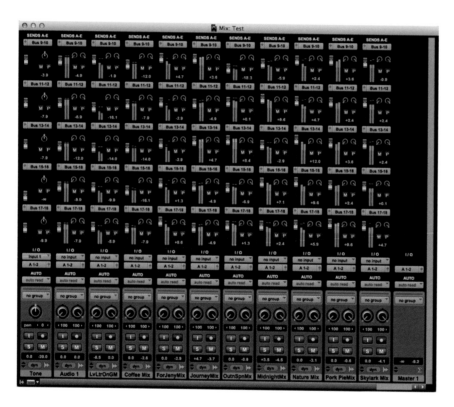

Figure 7.18
Expanded sends views with the main channel meters and faders hidden.

TIP

These expanded views can always be stored with window configurations, recallable via Memory Locations, to make it easy to change views as necessary. Storing an expanded sends view as a windows configuration lets you return to the normal view, using another configuration, very easily. All of these can, of course, be saved as a session template so you can have them all set up at the beginning of a session. See Figures 7.19, 7.20 and 7.21.

Figure 7.19
New window configuration for expanded sends view.

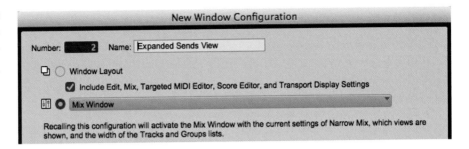

Figure 7.20
New Memory Location for normal view.

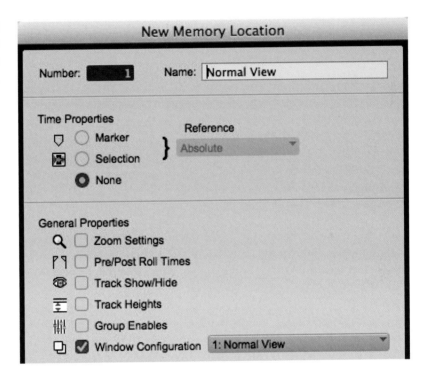

Figure 7.21
Memory Locations window set up to allow recall of normal and expanded sends views.

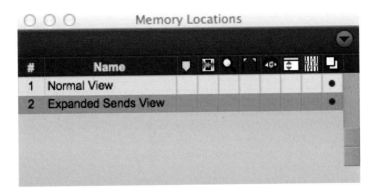

Metering and Loudness Discussion

To help you get a better understanding of some of the issues involved, here are some extracts from an article that I wrote for Pro Sound News Europe after discussions with Thomas Lund at TC Electronic and mastering engineer Bob Katz in the United States:

Thomas Lund has this advice for mixing engineers: 'If you apply the old-school sample peak detection methods of measuring level, peaks need to stay below −3 dBFS when delivery is linear, that is via CD, and below −5 dBFS when delivery is via MP3 at 128 kbps. That's not even always enough headroom – but 95% of the time it would be'. So, the message is that if you mix to digital, don't peak higher than −3 dBFS.

Lund also says 'Engineers should keep in mind that highly processed mixes with certain higher frequencies and high peak levels or clipping are likely to lead to levels in excess of 0 dBFS when reproduced at the outputs from typical D/A converters, which can lead to further distortion if there is insufficient headroom in the converters. Or whenever the signal crosses domains, such as in conversion to mp3'. Inter-sample peak metering to view and protect from these types of peaks is now available from TC Electronic in their System 6000 and from Sonnox and PSP in their peak limiters.

New tools are undoubtedly needed to help prevent the hyper-compression (squashed, loud mixes, and masters) arising as a consequence of super-high average levels and low peak-to-average ratios. Mastering 'guru' Bob Katz has plenty to say on this subject: 'The invention of digital audio started the accelerated loudness race which raised average levels on commercial CD releases almost 20 dB in 20 years. This was caused by the new ability in digital audio to normalize to the peak level. We need to return to the concept of headroom and standardize on an entirely new type of meter that is calibrated to true loudness and which allows for adequate headroom, with the true peak level hidden from the user'.

What Katz is referring to here is the situation that peak-only metering, especially in software DAWs such as Pro Tools and in typical CD players and other digital equipment, is no longer 'fit-for-purpose'. Katz's proposed alternative, the K-System, incorporates RMS metering and is coordinated to a calibrated monitor gain. RMS metering is more accurate than simple averaging, although not as accurate as a true loudness meter – which requires more DSP or CPU power and causes more latency.

According to Katz, mixing engineers would be better off dispensing with meters altogether and using their ears instead! As he explains, 'having calibrated monitor gain is just as important as metering peak-to-average ratios. It is possible to mix an entire album "blind", without any metering at all, yet never overloading the digital system! All you need to do is set a sufficiently high monitor gain (e.g. 83 dB at −20 dBFS RMS). When mixing this way, mix engineers can mix using their ears without the arbitrary constraints or influence of meters. The mixes which result will likely have a better crest factor (peak-to-average ratio) than typical mixes made while watching meters and, later on, in mastering, should produce louder masters with far less sonic compromise'.

As I pointed out, 'With the next (and even some current) generations of digital audio broadcast and consumer replay equipment, listeners will be given options to select dynamic ranges for replay. Consequently, music that has been squashed dynamically to sound louder will not sound louder any more, but will sound harsher in comparison with music that has not been so badly "mangled"'.

More of Bob Katz's views on metering are explained in documentation for the TC Electronic Finalizer, which can be downloaded at:

http://www.tcelectronic.com/media/1018177/katz_1999_secret_mastering.pdf

An AES paper written by Thomas Lund that addresses these issues can be downloaded at:

http://www.tcelectronic.com/media/1018175/lund_2006_stop_counting_samples_aes121.pdf

Other articles of interest can be downloaded at:

http://www.tcelectronic.com/TechLibrary.asp

MasterMeter Oversampling Metering

It has become fashionable in the music industry for clients to demand the loudest possible mixes – the theory being that these will stand out among the crowd. The problem is that when levels are at or near 0 dB Full Scale for much or all of the time, it is almost certain that when these recordings are played back on consumer CD players, significant amounts of fatiguing distortion will occur when the digital audio is re-constructed into analogue audio. This can happen even when the highest samples are kept just below 0 dB FS, because the re-constructed analogue waveforms can rise higher in level in between these sample peak levels – and will then be clipped by the analogue playback equipment.

The important point to note here is that the sample values in digital audio do not represent the peak level values of an analogue waveform re-constructed from these samples – peak levels will frequently exceed these values. Also, if you make a test CD and play it back on a high-quality CD player in your recording studio, this may have high-quality analogue electronics with sufficient headroom to successfully play back levels that exceed 0 dB FS without producing audible distortion or fatiguing effects – so you may be seduced into believing that there is no problem with your mixes!

The standard meters in Pro Tools will not alert you to this situation – they just measure sample peaks. What you need is a meter that simulates what happens during the reconstruction of the analogue waveform and alerts you if clipping would happen after reconstruction. If you know where the clipping takes place and by how much, you can go to these places in your mix and adjust levels to avoid this clipping.

There is a plug-in that will do all this: the MasterMeter plug-in that is now included with all Pro Tools systems. This allows engineers to compare regular and inter-sample peaks over time and make appropriate adjustments to counteract any clipping due to inter-sample peaks. Using an over-sampled peak meter like this, which simulates the reconstruction filters used in digital-to-analogue converters, is an essential tool to use – especially with mixes that have been processed to be particularly loud or 'hot'.

Using the MasterMeter

The MasterMeter has two separate meters, one that shows the standard signal level and another that shows the over-sampled signal level. As the oversampling process can create levels above 0 dB, this meter shows an expanded scale from 0 dB to +6 dB. It is easy enough to use: simply insert the MasterMeter on a Master Fader track and play the entire session back from beginning to end to check your final mix.

You can easily keep an eye on these meters while you play the mix and if you see any clipped signals, just stop playing back immediately and look for ways to fix the problem. This is fine if there are just one or two 'overs' during the mix. But if there are a lot, you will find it quicker to let the whole mix play through then look at the historical list of events that is created in the two 'browsers' provided for signal clips and over-sampled clips. Because the timecode for each event is listed, you can quickly go to each location where there is a problem so that you can fix it.

The Signal Clip Events browser displays historical clip events and has columns that show the relevant timecode for the beginning and ending of each clip event. With stereo tracks, the first column shows L or R to indicate if the left or right channel has clipped. The Min and Max values in this browser will always be zero, unless the Clip level is set below zero. At the bottom of the browser, the Peak field displays the highest dB value of the audio signal received so far.

The Over-sampled Clip Events browser displays Over-sampled Clip Events historically. The amount of potential clipping in excess of 0 dB is also displayed. The columns show the timecode for the beginning and ending of each clip event, as well as the minimum and maximum clip values created after oversampling. With stereo tracks, the first column shows L or R to indicate if the left or right channel has clipped. At the bottom of the browser, the Peak field displays the highest dB value of the over-sampled audio received so far. The Events field below the browser shows the total number of clip events in the over-sampled audio signal.

In the accompanying screenshot, Figure 7.22, you can see what was displayed when I inserted the MasterMeter plug-in on a Master fader in a Pro Tools session

Figure 7.22 MasterMeter plug-in inserted on a Master Fader with an audio track playing back 'Valerie' from the Mark Ronson 'Version' album. This shows 83 oversampled clip events.

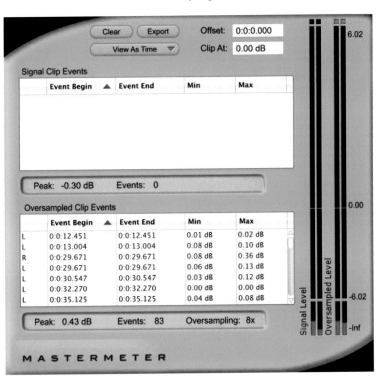

with just one audio track playing back 'Valerie' from the Mark Ronson 'Version' album. This shows the 83 Over-sampled Clip Events that occurred in either the left or the right or both channels and the times at which this clips occurred.

Pro Tools Faders

Audio Track Faders

When you play back an Audio track in Pro Tools, you are reading a file from a disk drive. If you set the fader at the unity gain position, neither boosting nor cutting the signal level, the meter (or meters on stereo or multi-channel tracks) should show a respectable signal level – not too low to have problems with the signal-to-noise ratio and not too high to cause clipping. Assuming that the audio was recorded correctly, this will be the case.

If the audio was recorded at too low a level, then you will probably hear background noise if you boost the fader a lot, and there will be little that you can do about this – which is why it is so important to get recording levels right in the first place. You may find it convenient to apply a gain change using AudioSuite to bring up the level of any tracks that are too low to avoid having to use ridiculously large boosts on these tracks with the faders. Pro Tools|HD faders will allow you to boost levels by up to 12 dB.

Master Faders

Master Faders control the output levels of output and bus paths. Post-fader inserts are provided, so you can also apply effects processing, along with the level control, for your main mix, headphone or cue mixes, and for any stems, effects sends, or other signal routing applications. Master Faders can also be used to meter a bus or hardware output to guard against clipping.

> **NOTE**
> Unlike inserts on Audio tracks and Auxiliary Inputs, Master Fader inserts are post-fader. So if you insert any plug-ins, you will affect the processing if you alter the Master Fader level, because changing the Master Fader level will change the level being sent to the plug-in. Also, don't forget to check that there is no clipping occurring in any plug-ins that you have inserted onto the Master Fader.

Although you do not have to use Master Faders in Pro Tools, you should normally choose to use these to get the best results. Master Faders do not use any DSP, so you won't suffer any performance 'hit' by using these, and, most importantly, Master Faders allow you to trim your final output level to avoid clipping the DAC (Digital-to-Analogue Converter) or 24-bit digital output when your mixed signals leave the Pro Tools mix environment. Also, by observing the Master Fader's meters, you can see if any clipping is taking place due to summing the individual channels.

If clipping is taking place, which will cause the Master Fader's red overload indicators to light up, you could lower all the individual channel faders to fix the problem, but it is much more convenient to simply lower the Master fader and play your mix again until you are sure that you have corrected the problem. The important point here is that you will not lose any quality by lowering the Master fader to adjust the gain at the output stage, because there is more than sufficient dynamic range available within Pro Tools 11 with its 64-bit processing capabilities.

Mixing Precision and Master Faders

The Pro Tools HDX hardware with Pro Tools HD software offers an expanded 64-bit floating-point mixer using 32-bit floating-point DSPs with large field-programmable gate arrays (FPGAs) to take care of inter-DSP communication at full 32-bit word lengths. This system retains 64-bit mixing values when mixers span DSP chips and uses the same mixer as host-based Pro Tools systems to make sure that mixes sound the same on both. It allows the use of a full 32-bit data path from audio file to final mix bus, obviating the need for many truncation and dithering points in the Pro Tools HD mixer.

Both host-based and HDX-based systems use dual-precision 32-bit (often called 64-bit) floating-point processing that provides much more headroom and dynamic range than the previous generation Pro Tools 24/48-bit TDM hardware. So a mix created on a Pro Tools HDX system can be transferred to a host-based Pro Tools system and it will sound the same. Any AAX DSP plug-ins will automatically be converted to AAX Native versions, and these will also sound the same.

This is a much better situation than previously, when the Pro Tools LE mixer would do a certain amount of work 'behind the scenes' to protect its presumed-consumer users from clipping bus summing points and plug-in inputs – so the mixes would require sometimes-extensive tweaking when transferred to Pro Tools HD systems that did not have these automatic 'protections' modifying the gain stages. Also, the RTAS plug-ins that used 32-bit floating-point calculations would sometimes sound different from their TDM 24/48-bit plug-ins, and in many cases, there were no equivalent versions available for both RTAS and TDM. With AAX, manufacturers are producing both DSP and Native versions at the same time which sound identical.

With 64-bit floating-point processing, you get far more mixer headroom than you will ever need – a theoretical 1680 dBs of dynamic range could be represented within this range of numbers! So, in practice, you will be able to mix as many tracks together as you like and you will not clip a bus input. Also, you can always use Master Faders on buses and outputs to scale the final results to a level that doesn't clip when the signal returns to 32-bit floating point in the Native/HDX mixer. So you can pull faders down to inaudible levels without losing any of the original signal resolution because the full 32-bit floating point 'word' is preserved all the way through the system. In practice, this means that it makes no difference whether you keep your channel faders low and have your Master fader high or have the channel faders high and the Master fader low – either way your recordings will retain their original resolution within the mix.

Typically, this is how it all works when you are using Pro Tools HD 11 (or 10) with an Avid HD Interface and an HDX card (as explained in more detail by Robert Campbell in his book Pro Tools 10 Advanced Music Production Techniques):

Digitized audio comes in from the A/D converters as 24-bit data with a 122 dB dynamic range from an Avid HD Interface. When analogue audio is converted to digital audio in the audio interface, it enters the mixer at a bit-depth matching the session settings, at 16-, 24-, or 32-bit floating-point. If the bit depth is anything less than 32-bit floating-point, the signal is padded with zeros to fill out the full 32-bit data path within the Pro Tools HDX system.

This is routed as 32-bit floating-point data to the plug-ins. AAX plug-ins can process the data at 32-bit, or at 64-bit double-precision if the plug-in developer has implemented this, shifting the 32-bit floating-point value to allow the 32-bit signal to enter a 64-bit plug-in without truncation. If this is the case, the data is then converted back to 32 bits before it leaves the DSP chip and is passed to the Mixer. The Mixer works at 64 bits throughout, with two 32-bit TDM channels used to transfer 64-bit data between mixer plug-ins on different DSP cards.

At the Mixer outputs, the signal is first converted from 64 bits (the two 32-bit 'words') to 32 bits. Master Faders let you choose which of these 32 bits reach the D/A converters where they are truncated to 24 bits, with 144 dB of dynamic range, before reaching the analogue outputs – with 125 dB of dynamic range available at the outputs of an Avid HD Interface.

As Avid's Simon Sherbourne points out, 'The key difference with the new mixer is that the signal path is 32-bit float throughout, so that the head-room and foot-room are maintained as the data passes through the plug-ins and busses. The Achilles heel of the older TDM systems was that signals were truncated to 24-bit at connection points, so it was easy to clip – and, unlike with the new system, clips were unrecoverable further down the signal path'.

So, although the 32-bit streams are summed using the 64-bit mixer, the interconnecting data paths remain at 32-bit floating-point, and a file recorded as 32 bits will pass through the entire mixing and processing path without any truncation or dithering. The combination of an FPGA chip to manage voices with floating point mixing and more powerful DSP means that the HDX cards have four times the processing power of the previous HD|Accel cards. And the 32-bit internal processing prevents internal clipping in plug-ins – which was always a danger in the previous generation HD and HD|Accel systems. The 64-bit summing allows for hundreds of tracks to be summed without clipping the HD software mixer or the input of a master bus, although the Master bus fader would need to be lowered accordingly to avoid clipping the outputs.

The 32-bit Advantage

So what are the lessons to be learned from all of this? First of all, you should start using the 32-bit floating-point capabilities as soon as possible – creating new sessions in this format and recording your audio to 32-bit floating-point files. You can use Pro Tools 11 with a powerful computer and get very good results, but for the best results you should use one or more Avid HDX cards so that these can take care of the AAX plug-in processing and mixing, leaving the CPU free to carry out additional processing – for example, running a large sample library from the Vienna Symphonc Library.

A major advantage of using 32-bit floating point files is that if you have bounced or recorded a file to disk and the audio has clipped, you can recover un-clipped audio from this! Just open the file in Pro Tools and pull down the Clip Gain, or reduce the gain using the AudioSuite Gain plug-in, until the clipping disappears – impossible to do with fixed-bit files!

Using a Master Fader to Prevent Clipping on a Bus

The way a Master fader works with the 64-bit floating-point mixer is to create a 32-bit 'window' into the 64-bit 'data space', changing the exponent of the 32-bit 'word' to raise or lower the output level.

A Master Fader can be used to control the level of the audio on an output or outputs. It can also be used to control the level on a bus. This is actually a very important application of Master Faders that deserves to be more widely used.

For example, it is very common to sub-mix drums onto an Auxiliary Input channel, so that you can control the overall level of the kit using one fader and so that you can insert effects such as compression or reverb across the whole kit – see Figure 7.23. The problem here is that sometimes the combined signals from all the Audio tracks that are feeding the Auxiliary Input can cause the level on the bus to clip, even though the levels of individual Audio tracks do not clip. And you can't use the fader on the Auxiliary Input (or on a Send) to prevent this clipping.

Figure 7.23
A typical sub-mix of drum tracks onto an Auxiliary Input with a Send used to add reverb to the whole kit. Note that the clip lights are yellow on the Auxiliary Input and on the reverb Send.

As you can see in the accompanying screenshot, Fig. 7.24, the clip lights are yellow on the Aux Input and on the Send to the reverb. Lowering the Auxiliary Input fader or lowering the Send fader will not prevent this clipping, as these faders simply control the amount of the clipped audio that is sent to the mix bus and to the reverb device. If you try doing this, the meters will still show that the audio on the bus is clipped.

Figure 7.24
Using a Master Fader to monitor levels on the bus used to sub-mix the drums onto an Auxiliary Input. Note that the levels on the bus are clipped, as revealed on the Bus Monitor meters and on the Auxiliary Input meters and the Send meters. Lowering the Send level will not prevent clipping, and lowering the Aux fader level will not either.

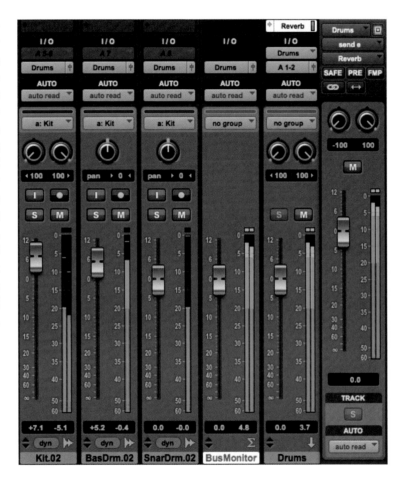

You could lower some or all of the individual faders until the summed level at the Auxiliary Input no longer clips, but you run some risk of disturbing the balance you have carefully set up between the individual tracks, and this is not as convenient as lowering one fader.

Fortunately, there is a solution to this problem – using a Master Fader to monitor and control the bus levels. Simply insert a Master Fader and assign this to the bus that you are using to route the Audio tracks into the Auxiliary Input to form your sub-group.

Whenever you see the clip light show on a Master Fader that you are using as a Bus Monitor, you can lower the level of the Master Fader to reduce the level on the bus until it stops clipping. You could also insert a compressor or limiter on the Master Fader to 'tame' the peaks – see Figure 7.25.

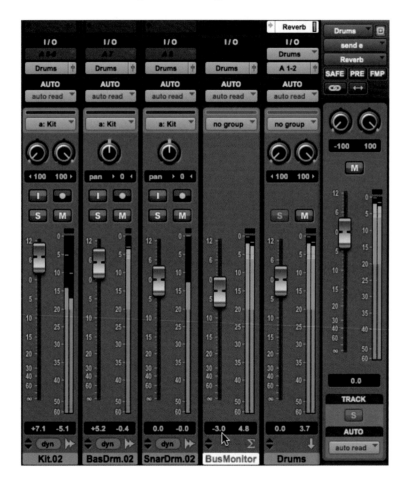

Figure 7.25
You can lower the Master Fader used to monitor levels on the drums sub-mix bus to prevent clipping, as has been done here.

Workflow Enhancements for Pro Tools 11

Bypass Inserts

There are eight new key commands to allow you to bypass inserts that work in conjunction with whichever tracks are currently selected: Shift-A bypasses all inserts, Shift-1 bypasses inserts A-E, and Shift-2 bypasses inserts F-J.

These allow you to check what your mix sounds like with all processing bypassed, or you could split your processing into two different sections of inserts so that you can make A/B comparisons of different approaches to processing your mix.

Bypass Effects Categories

You can also bypass the different effects categories: Shift-E bypasses all EQ's, Shift-C bypasses all dynamics, Shift-V bypasses all reverbs, Shift-D bypasses all delays, and Shift-W bypasses all modulation plug-ins. Again, this is a very useful feature when you are setting up your mix sessions.

> **NOTE**
> When EQ or Dynamics plug-ins are bypassed on channel strip plug-ins, the inserts turn purple (see Figure 7.26) to indicate that a process within the channel strip has been bypassed, but not the entire plug-in. This only works in plug-ins that support this feature, such as the Avid Channel Strip.

Figure 7.26
Channel Strip plug-in shown in purple to indicate a bypassed process.

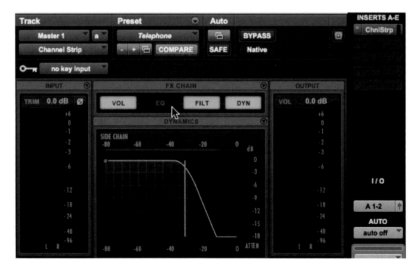

Shortcut to Plug-in Presets

Another handy new feature lets you right-click on a plug-in to select presets without opening the plug-in – see Figure 7.27.

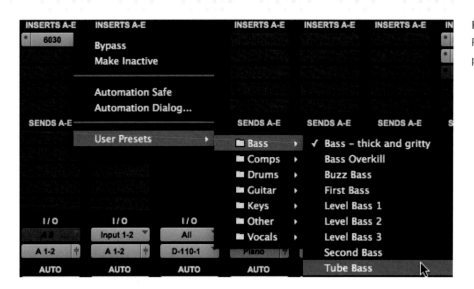

Figure 7.27
Right-Click shortcut to
plug-in presets.

TIP

Control-clicking (Mac) or Start-clicking (Windows) will bypass that insert and all inserts after it. There are similar key commands for Sends. Shift-Q will mute all sends, Shift-3 will mute sends A-E, and Shift-4 will mute sends F-J. You can also Control-click (Mac) or Start-click (Windows) on any send to mute that send and all sends after it.

Restore Previously Shown Tracks

You can now restore previously show tracks from the Tracks menu. Say you had a particular set of tracks visible and wanted to select a new selection of tracks to view. You would go to the Tracks menu from the Tracks list and choose Show Only Selected Tracks – see Figure 7.28.

Now to go back to the previous view, go back to the Tracks View menu and choose Restore Previously Shown Tracks – see Figure 7.29. This lets you quickly go to a custom view to another view and back without having to save a track hide/show locator first.

Figure 7.28
Show only selected
tracks.

Figure 7.29
Restore previously
shown tracks.

The 'Tools of the Trade'

Mixing primarily involves adjusting the levels of the various instruments, voices, and other mix 'elements' such as sound effects that make up your mix. Beyond simple level control, there are many other processes that may be involved in creating a finished mix. Pro Tools provides you with, or lets you access, all the 'tools of the trade' that you will need to make even the most complex mixes. These 'tools' include the wide selection of signal processing plug-ins supplied with your Pro Tools system, third-party plug-ins that you can buy, and any external hardware effects that you hook up to your system.

Level Balancing

If you come across older books about sound studios, or if you work for the BBC, you will hear the term 'balancing' used to describe what most people call 'mixing'. 'Balancing' is actually a very good way of describing the most important thing that you will do when you are mixing a multi-track recording. This is to decide how loud or soft the various elements of the mix should be – which you do by 'balancing' the level of each instrument or sound against the others, moving each up or down in level until the mix sounds the way you want it to.

> **TIP**
> The balance in levels between the various mix elements is probably the single most important aspect of most mixes. Get this right and you are a long way along the path toward a great mix!

One decision to make is whether to keep the level balances similar to those that you would expect to hear in a 'live' situation or whether to balance the instruments in such a way as to create unique perspectives that would not normally be heard in 'live' playing situations.

If you were recording a string quartet playing conventional repertoire, for example, you would record this in a suitable room (in a recording studio or at a 'live' venue) in stereo to capture the overall effect, possibly with individual microphones on each instrument so that the balance between these could be fine-tuned at the mixing session. A jazz group could be recorded in a similar way, with the aim of delivering the most realistic, 'believable' sounds to the listener. The recording engineer would aim to capture the sound of the instruments arranged as they typically would be onstage and as heard by a member of the audience located centrally in the listening area.

With many genres of popular music, on the other hand, it is more likely that you will create balances between the instruments that you would not normally hear when these instruments play together in a room. For example, you might have the drums very loud in the mix with a brass section playing riffs much more quietly in the background. Or you might have a quiet acoustic instrument such as the ukelele turned up loud enough in the mix to be heard playing a solo above a full electric band with keyboards, guitars, bass, drums, and so forth all playing at the same time. Or maybe you could have some percussion instruments such as a triangle or tambourine playing at a very low level in the mix – but just sufficiently loud to be heard and have a useful effect.

Panorama (Pan) Positions

Positioning the mix elements within the stereo 'panorama' is another important part of the 'art' of mixing. Panning is the subject of much debate and ultimately comes down to your own creative choices. To create a satisfying mix, your goal should be to arrange the positions of the mix elements within the stereo sound field such that they work well together without obscuring each other. It is also a good idea to aim for some kind of symmetry so that you don't end up with lots more going on to the right or to the left. You might pan one percussion instrument to the left and another to the right, for example. Or you might have an acoustic piano on the left with organ or electric piano on the right. Of course, there will be times when the way the musical arrangement develops means that this is not always possible, but it is generally a good goal to aim for.

> **TIP**
> If you don't have another instrument in the arrangement that you can use to create balance, you might pan a delayed version of one of the mix elements to the other side. This can be very effective with guitar or percussion parts, for instance.

Some early stereo mixes used radical pan positions, such as bass and drums hard left, keyboards and guitars, brass or percussion hard right, and vocals in the centre. This was sometimes due to technical limitations in the equipment used. On other occasions this was done as a result of the creative choices made by the engineers and producers. Mixing this way has the advantage of leaving lots of room for the vocal to be the absolute centre of attention with the other instruments very much in a supporting role – which is ideal if you have a great singer and a great song.

I have recently been listening to some early stereo mixes of late '60s pop/soul songs – hit records such as 'Goin' Out of my Head' by Little Anthony & The Imperials and 'Hey Girl' by Freddie Scott – which typically might have the rhythm section panned hard left, backing vocals, brass, and strings panned hard right and lead vocals in the centre. This way of mixing had the advantage that the lead vocal was easy to hear – loud and clear and with no competition from other instruments occupying this central space. This technique is still occasionally used today. One example of this is a song called 'Empty' on Ray Lamontagne's 2006 album 'Till The Sun Turns Black', which has drums panned left and acoustic guitar panned right with the vocals in the centre.

Nevertheless, it is much more usual today to have the lead vocal, the bass guitar, and the bass drum all dead centre, with the snare sometimes centre or maybe, as with the hi-hat, panned left or right, and with the kit overheads capturing a natural stereo spread of all the drums including the cymbals. Talking about panning the drums, you may wish to consider whether to pan these the way the audience would hear them or as the drummer would hear them. There are arguments in favour of both ways. If you pan from the audience perspective, this is often what the audience expects and probably sounds the most natural, while panning from the drummer's perspective lets you listen from the drummer's seat on-stage.

Guitars might typically be panned half left or right, with keyboards spread a little further out – but don't be afraid to experiment here. It can sometimes be very effective to use an auto-panner (or Pro Tools automation) to move instruments around as the music plays – like a wandering percussionist or acoustic guitarist might do on-stage.

It is also worth keeping in mind where the mixes are going to be played. If you are aiming them at the audiophile with a great stereo system listening in the sweet spot, then you can make good use of extremes of panning. On the other hand, if the mixes are intended for a dance club, even if this has some kind of stereo speaker system installed, you can't count on dancers being in any kind of sweet spot for too long – so you should keep the important rhythmic elements of your mix much closer to the centre.

Although the stereo panorama that you can create using just two speakers basically allows you to move mix elements along a horizontal plane running from far left to far right, it is also possible to create a sense of depth, and even of height, using careful choices of delays and reverberation. A word of warning here: you do need to be careful about your choices of pan positioning if you are aiming for good mono compatibility. If you have two guitar parts playing in similar frequency ranges, for example, you can make these clearly audible as individual parts in a stereo mix by panning them hard left and hard right. Unfortunately, when you listen in mono, it will often be much more difficult to distinguish one part from the other as they will both be heard playing in the same place.

Forewarned about this (by having checked in mono), you can try other ways of distinguishing the two parts, perhaps by EQ-ing the two parts differently and then narrowing the distance between the two parts in the stereo mix (so that there will be less of a difference when you 'collapse' from stereo to mono) before checking again in mono to see how well this works.

> **NOTE**
> If you want to create a mix with good mono compatibility, you should regularly check how the mix sounds in mono and adjust the mix accordingly.

> **TIP**
> If you pan mix elements while monitoring in mono, you may be able to find an optimum pan position for each element in which it sounds more audible when summed to mono.

Of course, if you are working in surround, you will have many more choices when it comes to panning: 360 degrees around the listening position and all points within. This is why surround mixes can sound much more natural, because it is much easier to achieve separation between the different instruments in the mix.

Panning Synthesized and Sampled Sounds

By default, most synthesizer and sampler sounds are set up to work in stereo, panned hard left and hard right. One thing to watch out for is the trap of including these synthesizer parts in stereo in your mix – typically with the dry sound panned hard left and a chorused version panned hard right, although sometimes the second channel contains synthesized sound components as well. You should almost always reduce the width of the panning and place these sounds more appropriately within your mix. Better still, just use the original synthesized or sampled sound in mono (or stereo) without its reverb or other effects, so that you can position this exactly as you wish in your mix and apply the higher-quality reverb and other effects that you have available in Pro Tools. Your goal should be to find pan positions for the synthesized instruments that work in the context of your mix – and the sound designers who set these sounds up for your synthesizer or sampler could not have panned these suitably except by lucky accident. So don't go with the defaults – pan the synthesizers carefully yourself!

Pan Laws

Variable stereo pan depths are available in Pro Tools from the Session Setup window. The default setting for this is −3dB in Pro Tools (it was −2.5 dB in Pro Tools 8 and previous versions) and it can be adjusted from −2.5 to −6 dB.

As Avid explains: 'There are a variety of ways that analogue and digital mixers handle *stereo pan law*, the level calibration when a signal is set to centre. For example, a track set at unity gain that is panned centre will output a level of −3 dB to both the left and right speakers. But when you pan it hard left, it will output at unity gain'.

This 'pan law' as it is normally termed compensates for the 3 dB boost that happens when panning sounds from the extremes to the centre position by subtracting 3 dB from the level in the centre. This keeps the level of the panned elements the same when they are moved from the extremes to the centre position.

When you move a sound into the centre position, it is more likely to be masked by the many other sounds that are likely to be positioned in the centre, so you may prefer to use the −2.5 dB setting, which will boost the level when panned centre by 0.5 dB.

The pan law used in many mixing consoles, particularly for surround sound, is −3 dB, so the new default setting generally makes much more sense by falling in line with this industry standard.

The pan law used in many older UK-designed analogue mixing consoles is −4.5 dB for stereo or −6 dB for full mono compatibility, so it is good to see that these options are now available for Pro Tools – see Figure 7.30.

Figure 7.30
Session Setup window showing Pan Depth options.

Using Compressors to Smooth Out Level Changes

Once you have panned your mix elements and achieved a rough balance, you should listen to the whole piece of music from beginning to end. If you notice that any of your tracks are uneven in level throughout the piece, you can always use the volume automation in Pro Tools to smooth out these level changes.

Of course, this can become a very time-consuming task if the singer, guitarist, bass player, or whoever, is not experienced enough to sing or play very evenly, so that their levels are constantly changing throughout the piece. You can often get quicker results by using compression to smooth out the level changes, then setting the overall level for the track using a combination of the makeup gain in the compressor and the fader level in the mix.

Just to remind you here, a compressor reduces the highest peaks in the audio, making the sound more even. When the peaks have been reduced, you can then raise the average level of the audio without the peaks exceeding the maximum level that the system can handle without distortion.

TIP

You may not need to use compression on the bass, guitar, drums, or other instruments if you are recording experienced studio musicians. I have often been amazed at how evenly the top studio musicians here in London are able to play – with wonderful control over the dynamic levels that they are producing from their instruments.

NOTE

Compression should always be applied carefully: it is all too easy take away the natural 'openness' of a recording when you apply too much compression – 'squeezing' the 'life' out of the sound. And when you have chosen the compression settings that you think are working for you, be sure to compare the sound without the compression again to make sure that you are really happy with the results.

Using the Gain Plug-in to Smooth Out Level Changes

Another way to smooth out unwanted level changes is to select any audio that is too loud and apply a gain change using the AudioSuite Gain plug-in – see Figure 7.31. This way, you avoid all the unwanted side effects of compression, such as the 'breathing' effect that you sometimes hear when the compressor kicks in and then releases. It is very easy to do this for just one or two notes or sounds that are too loud and this can be a useful technique to use in combination with other techniques for smoothing out the levels.

Figure 7.31
AudioSuite Gain plug-in.

De-essing

Another problem that often becomes apparent during a mix session is sibilance on the vocals that becomes even more noticeable when you raise the level and maybe boost the mid and upper frequencies using EQ to make the words easier to hear. In this case, it is just the so-called 'sibilant frequencies' – higher frequencies that are prominent in the syllables 's' and 'f' – that need to be compressed because they are too loud.

The trick here is to use a 'de-esser'. This is a compressor designed specifically to alleviate problems with Ss and other sibilant sounds by applying fast-acting compression just to the sibilant frequencies. Such a device uses EQ filters in a 'side-chain' to isolate the sibilant frequencies so that the compressor will only affect the levels of the problematic frequencies, leaving the rest unaltered.

> **TIP**
>
> As when applying level changes to balance overall track volume levels, it is also possible to apply individual gain changes just to the sung notes that are suffering from sibilance problems. And, again, you can apply these changes in real-time using Pro Tools volume automation, or as an 'off-line' process using the AudioSuite Gain plug-in. If there are just a few sibilant notes that need adjusting, or if you have a lot of patience, this can be a more effective technique to use than using a de-esser – which is a more automatic process that does not pay such detailed individual attention to each sibilant note.

Creating Dynamics within the Mix

Often you will realize that some of the mix elements ought to get softer or louder during the mix to create and maintain interest. For instance, if you have a solo instrument in the mix such as a guitar player who improvises throughout the piece, you could turn this up louder wherever nothing much else is happening during the music and make it quieter whenever there are lots of other mix elements vying for the listener's attention.

Also, some of the constant mix elements could be muted in particular sections so that they make the sections where they can be heard stand out better. So, for example, you might record a guitar or keyboard playing a rhythmic figure throughout the piece, then try muting this in the verses – just using it in the choruses.

> **TIP**
>
> Don't hesitate to record or add new material during the mix session if you feel that something is missing musically from the arrangement. When the singer is not singing, for example, you might want a guitar or sax to play a solo that was not thought of during the track-laying sessions but you now realize that this would be the 'icing on the cake' to complete the recording.

Frequency Balancing

You can balance the frequencies in a recording either by the choices that are made in the musical arrangement or by using the tools available in the recording equipment that you are using.

Musically

Having different instruments play in different frequency ranges (i.e. in different octaves) usually helps to make a more satisfying sound than if several are playing in the same range, where they may clash with each other. This choice of which instrument plays where and when may be made by the musicians or bandleader, or it may fall to the songwriter, composer, arranger, or producer.

You should always aim to create a good frequency balance in the musical arrangement, which is why it can really help to employ the services of a good musical arranger to work on your music production if you do not already have someone with these skills involved.

Technically

You can also use EQ to adjust the frequency balance of any of your mix elements, so that you create a more balanced range of frequencies within your mix. For instance, electric or acoustic guitars can have quite a lot of low frequency content that can interfere with the bass guitar or other low frequency instruments. Filtering out the lower frequencies or reducing the level of the low frequency content can make the guitar parts blend into the mix much better.

For example, I wanted to blend in a recording of a Martin HD28 acoustic guitar with a mix that also contained bass guitar, drums, electric guitar, tenor banjo, mandolin banjo, vibes, percussion, and vocals. I used the following EQ settings on the Martin guitar to achieve the result I was seeking:

> Subtracting (−) 3.5 dB at 50 Hz took some of the 'muddiness' away from the bottom end, allowing the bass guitar and bass drum to be heard with less interference from the acoustic guitar's low frequencies.

> Adding (+) 1.5 dB at 100 Hz gave the guitar more 'body' or 'fullness', helping the sustaining chords to act as a 'pad' sound.

> Adding (+) 1.5 dB at 3.2 kHz 'brightened up' the chords in the mid-range, making the tonal detail and the pick strokes much easier to hear.

> Adding (+) 1.5 dB at 6.3 kHz let the guitar 'cut through' the mix with much more clarity.

Remember that this is just an example. Other than by coincidence, these settings are not going to work for you because the guitar will be different, the player and pick will be different, the microphone and preamplifier will probably be different, the room will be different, the other instruments in the mix will be different, and so forth. So you will always have to find your own settings by trial and error – although you could always try the above settings as a starting point and you just might get lucky!

One way to look at EQ is to regard the boost/cut controls for the various frequency bands simply as volume controls for those bands – which is, essentially, what they are. When you think of EQ this way, it is easy to see that lowering the volume of particular frequency bands allows those that are not cut to stand out more prominently. So instead of boosting the mid-range you could simply cut the highs and lows and the mid-range will be emphasized. Conversely, if you want to emphasize the highs and lows you can cut the mid-range. This approach also helps to avoid causing clipping, which can happen when you raise levels.

Reverb

Reverberation occurs naturally in rooms, halls, and most other acoustic spaces and is composed of myriads of reflected sounds bouncing around the acoustic space, reflected by walls, floors, ceilings, and any other hard, reflective surfaces, such as glass windows or screens within the space. Sound is absorbed by soft furnishings, curtains, carpets, human bodies, and so forth, so in real rooms, the reverberation will change according to how many people are present, and even whether the curtains are drawn or not. Harder surfaces reflect sound, and if there are many small surfaces that reflect the sound in different directions, these will make the reverberant sounds more diffuse. To make a small room sound larger than it is, it is a good idea to use special diffusers.

In classical, orchestral, and some jazz and other recordings, the goal is to capture the natural reverberation of the room along with the sound of the instruments. So-called 'ambience' microphones can be placed wherever is appropriate in the room to capture this sound and blend it in with the mix. Artificial reverberation is often added to recordings to enhance the natural reverberation that already exists or to create reverberation around directly injected electric or electronic instruments such as guitars or electric pianos, or around synthesized or sampled sounds.

My first choice of reverb for Pro Tools systems has to be Altiverb from Audio Ease. This is a convolution reverb that has the largest set of high-quality impulse responses, taken from many of the world's finest rooms, halls, chambers, studios, and so forth, giving you access to some of the best acoustic

> **TIP**
> On a vocal, you might use a long pre-delay setting with the reverb, so that the words retain lots of punch and audibility, while gaps between words are enhanced by the sound of the reverb kicking in.

spaces from around the world. I always use this as my primary reverb. I have compared all the other reverb plug-ins for Pro Tools systems, and Altiverb is way ahead of everything else. It was the first convolution reverb available for Pro Tools systems and has reached a stage of development that none of its competitors have been able to match.

Avid's D-Verb, provided with all Pro Tools systems, is very basic, but provides a good range of reverb types, including Hall, Church, Plate, Room 1 and 2, Ambient, and Non-linear. Avid also offers its more advanced Reverb One plug-in as an option. Reverb One has a variety of controls for producing a wide range of reverb effects and also has a Dynamics section that can be used to modify a reverb's decay character, making it sound more natural, or conversely, more unnatural, depending on the desired effect. A third option from Avid is ReVibe II, which provides higher-quality acoustic environment modelling with plenty of controls for Early Reflections, Room Coloration, and Chorus effects and works in all the surround formats supported by Pro Tools.

Delay Effects

Delays can be used to produce obvious effects such as echoes, plus many other effects. For example, you can create a stereo image from a mono sound source using a short delay and set the width of this image by panning the delayed sound away from the source sound to create the required distance between them. If the delay is short enough, the two sounds will be heard as one sound with a wider image than the original source.

One thing to watch out for here is that if the mix is summed to mono for playback on an old radio or TV or whatever, the image widening effect, of course, disappears. So if you have been relying on having particular instruments widely panned to prevent them from obscuring other elements within the mix, then when you sum to mono your mix may suffer badly as a result. To counteract this tendency, make sure that you don't pan the mix elements too widely. Obviously, the closer you pan mix elements to the centre, the better the transfer to mono will be.

Noise Reduction

Sometimes, when you bring the level of a particular track up as loud as you want it to be in your mix, you will hear some background hiss (e.g. from a guitar amplifier) that you definitely don't want to hear. This is where you might want to use noise reduction software (or hardware) to get rid of the hiss (or other noises).

I often record a Wurlitzer electric piano that has a very noisy preamplifier. This produces lots of hiss and 50-cycle mains hum that can often be very intrusive, especially on quiet passages, so I often find myself de-noising Wurlitzer tracks.

Noise Gates and Expanders

An expander expands dynamic range rather than compressing it, so expanders can be used to reduce signal levels when the input signal falls below the selected threshold – making the difference between softest and loudest signals greater. This is called 'downward expansion' and is just what you need to prevent any low-level noise that occurs between 'wanted' audio (with levels above the threshold) from being heard. This is why an expander can be used as a so-called 'noise gate'. When the input signal falls below the threshold, the expander makes the quiet sounds so quiet that they cannot be heard, that is, the gate closes. When the audio input signal rises above the threshold again, the gate opens.

TIP

I find myself hardly ever using noise gates within Pro Tools because it is so easy to define a 'silence' threshold and strip out any audio that falls below this threshold. Doing it this way avoids any possibility of false triggering where an unwanted sound just above the threshold opens the gate. Of course, you still have to listen and check carefully to make sure that what you have defined as 'silence' has stripped out everything you want to remove correctly.

Pitch and Time Correction and Manipulation

Pro Tools offers an AudioSuite plug-in called Time Shift that can be used to shift the pitch or time (speed/length) of any audio clip, changing the length of a clip without changing its pitch or changing the pitch without changing the length.

For example, the guitarist might have played just one wrong note, say, a semitone lower than the correct note. Using a pitch shifter, you can select just this one note and move it to the correct pitch without having to trouble the guitarist to come back and correct this. Or maybe the mandolin player played an extra note in between two important melody notes and you have edited out the unwanted note but the first note now doesn't sustain as long as it ought to. Just select the sustain portion of this note and time-stretch it to fill the gap from which the unwanted note was removed.

A Varispeed mode is also provided. In this mode, you can adjust either the Time Shift or Pitch Shift control to set the amount of time and pitch change in terms of a percentage of the original. In this mode, when you increase the pitch of a clip, the speed of the clip also increases and its length is shortened – and vice versa.

TIP

If you want to shift the pitch of vocals, or any instruments with pronounced formant frequencies, then you should enable the Formant controls and make sure that the formant frequencies have a suitable amount of pitchshift applied – a zero amount would leave the formants untouched as you shift the pitch of the vocals, for example.

Time Shift works well enough for small shifts of just a few semitones or beats per minute, but undesirable artefacts quickly appear with greater shifts. Avid also offers the more advanced X-Form plug-in. This offers the highest quality pitch and time shifting algorithms and lets you alter audio files by even-larger percentages with even fewer undesirable artefacts.

Vari-Fi

Vari-Fi is an AudioSuite plug-in that provides a pitch-change effect similar to a tape deck or record turntable speeding up from or slowing down to a complete stop – see Figure 7.32. Vari-Fi preserves the original duration of the audio selection.

There are two 'Change' control options: if you select 'Slow Down', this applies a pitch-change effect to the selected audio, similar to a tape recorder or record turntable slowing down from normal speed to a complete stop. If you select 'Speed Up', this applies a pitch-change effect to the selected audio, similar to a tape recorder or record turntable speeding up from a complete stop to normal speed. This effect does not change the duration of the audio selection.

The 'Selection' setting determines the duration of the rendered clip in relation to the processing. When the 'Fit To' option is selected, the length of the audio

Figure 7.32
Vari-Fi.

selection is retained when rendering the AudioSuite effect. This is useful for rendering the effect in place (especially if the selection is constrained by the grid or by adjacent clips). When this option is enabled, processing is applied to only two-thirds of the selection so that the resultant rendering maintains the original duration of the selection.

When the 'Extend' option is selected, all audio in the current Edit selection is processed and rendered. The resulting rendering is 150% the duration of the Edit selection. The selection start point does not change, but the rendered clip extends beyond the end of the Edit selection. This can be useful if the last third (for speeding up) or the first third (for slowing down) of the Edit selection needs to be heard in the rendered effect.

There are two 'Fades' controls: If you select 'On', a fadeout will be applied if Slow Down is selected or a fade-in will be applied if Speed Up is selected. When 'Off' is selected, no fade-in or fadeout is applied when you render the Edit selection. This can result in a more pronounced 'tape-stop' or 'tape-start' effect and can also be useful for preserving the dynamic level at the end of the Edit selection when Slow Down is selected, or at the beginning of the selection when Speed Up is selected.

Automatic Pitch Correction

If the vocals, or any instrument that is playing single notes (as opposed to double notes or chords), are out of tune, there are various specialized software tools available from other manufacturers to correct these. Auto-Tune is the best-known pitch correction software for Pro Tools. This has an automatic correction mode that is very easy to use and a manual, graphical mode for more advanced corrections. Auto-Tune itself is now being rivaled by Melodyne, which has a more comprehensive feature set than Auto-Tune and can handle multiple tracks of audio. Melodyne's new Direct Note Access technology will even allow you to access individual notes in chords and polyphonic audio – letting you see them, grab them, and edit them!

Other Effects

There are times when you will want to use effects such as vocoders, harmony processors, or other unusual effects processors. Or you might like to work with a multi-effects processors. Early examples of these would include Yamaha's

SPX90 and Eventide's H8000, which included reverbs, delays, tremolo, chorus, flange, and so forth. The Eventide H3000 Factory multi-effects processor, now available as an AAX plug-in, features the most comprehensive set of modulation effects that I have come across. Sound Toys Phase Mistress and Filter Freak plug-ins provide tempo synchronized modulation effects with lots of great presets for, respectively, phase and filter effects. But my favourite multi-effects processor has to be Sound Toys Crystallizer. As it says on the website: 'Inspired by the Crystal Echoes preset in the Eventide H3000, Crystallizer combines granular reverse echo slicing, and retro pitch processing to create a huge range of radical sonic manipulations and classics with a twist. Use it to create synth-like textures from simple acoustic guitar rhythms, lush detuned echoes, or completely psychedelic pitch-shifted reverse echo effects. Great for drums, guitar, bass, sound design, electronic music, and just about anything else – Crystallizer is a truly unique and creative effects processor'.

Re-amping and Amplifier Simulators

If you have the equipment available and have the time and energy for this, you can always route your guitar (or any other) track out from your Pro Tools interface into a real guitar amplifier in your studio, put a microphone in front of its loudspeaker cabinet, and record this back into Pro Tools. This is called 're-amping' for obvious reasons. You can even go nuts and hire in lots of tasty vintage gear for the session – which costs a lot less than owning all the gear! But this can still be an expensive business both in the studio time needed and the costs of hiring in the gear.

There are various AAX plug-ins now available that emulate guitar amplifiers, speakers, effects, and even the microphones to record these! For example, Softube offers its Vintage Amp Room, which faithfully recreates just three great guitar amps in a complete studio setup with speaker cabinets and fully flexible microphone positioning. Vintage Amp Room has no extra gadgets or weird-sounding presets – it just accurately simulates classic Fender, Marshall, and Vox setups.

Avid offers its own re-amping tool, called 'Eleven' – see Figure 7.33. This is also available in hardware as a rackmountable unit. Eleven has a great selection of Fender, Vox, Marshall, Soldano, and Mesa Boogie amp models plus convolution-based speaker combinations and mic modelling.

Figure 7.33
Eleven.

All these plug-ins are great for 're-amping' electric guitars, basses, or keyboards, so that you can change the sound of these during your mix session. You can simply directly inject your electric instruments into Pro Tools while recording, then build the sound you want for your mix at your leisure later.

> **NOTE**
>
> You can always change the sound of an instrument that was previously recorded through an amplifier or effects, but it works best if you have recorded a relatively clean sound with minimal effects if you want to get the best results when re-amping and adding new effects.

It is always better if the guitar player plays through the amp that will be used for the track so that he or she can respond as a player to the sound of the amp. If it is a clean sound, the player will usually adjust his or her playing according to the sounds he or she is hearing from the amp. With a distorted sound the player usually plays differently – may be hitting more harmonics with the pick or holding the notes longer because they sustain longer with the distorted sound on the amp. And the way the player digs into the strings with the pick has a big effect on the way the amp responds. If the player uses thumb or fingers, the sound changes again, with the amp responding differently to the way the strings are driving the input.

Working on Your Mix Session

It can be a good idea to let some time pass between your original track-laying sessions and the mix session. When you come back to it with 'fresh ears' you may hear things from new perspectives that can lead you to making a much more interesting mix than you had at first imagined.

But it can go the other way as well: you sometimes get the best mix by building everything up while you are recording, making edits or adding effects as you go along, then tweaking and balancing everything to make the final mix as soon after the recording phase is finished as you are ready.

'Housework'

One of the first things you will probably do at the start of a Pro Tools mixing session is to listen through to the tracks to see what's there – especially if you are mixing music for a client that you have not heard before, or if it is something you are coming back to from some time ago.

This is a good time to be doing some 'housework'. You could start by making sure that you have a separate backup copy before you do anything, so that if you mess up in any way you can quickly get back to the original files. Then you should delete anything that you are certain that you will not need – secure in the knowledge that you can always restore the session from your backups if you make a mistake.

There might be lots of MIDI tracks that have since been replaced by real instruments, or vice versa. It is also quite likely that there might be several 'takes' of the lead vocal or of a guitar or sax solo that may need to be 'comped' to make one or more composite tracks containing the best phrases. There might be various layers of tracked-up backing vocals, rhythm guitars, or brass parts that could be mixed down to stereo to form more manageable sub-groups or 'stems'.

There are probably lots of unused clips that you could easily get rid of to make it easier to see the ones you do want, although you will need to take care before deleting any whole clips to make sure that clips within these are not needed anywhere in the session. Again, if you make a mistake, you have your backups.

Track Ordering

You will probably want to establish the order of the tracks in the Pro Tools Mix and Edit windows the way you prefer this to be. I always have any Master tracks at the right of the mixer, with a click at the far left, drums next to this, followed by bass, keyboards, guitars, any other instruments, then vocals. I know other people who would reverse this order. The main idea here is to have some kind of logical ordering of the tracks that makes it easier and quicker for you to work with these.

It can also be helpful to name some (or even all) of the clips, takes, or tracks more appropriately – especially if this will make things easier and more efficient for you when you get to the creative part of the mixing session. Don't forget that you can easily save alternative versions of any Pro Tools session if you want to keep older versions in case you need to revert to these.

Removing Unwanted Sounds

It is always a good idea to take note of and then mute or delete any odd noises or sounds, clicks or pops, microphone stand rumble, people talking in the background (it happens), excessive bleed between microphones, or any sounds that you don't want to appear inadvertently in your mixes. You may have to solo tracks one at a time to find these unwanted sounds, but this is not really too much time wasted (as long as there is not too much of this stuff going on) if you view it as a useful part of the process of familiarization that most people need to go through before getting creative with the mix (because you get to hear what is really happening on the tracks that you solo).

Noise gates can be used to reduce or remove bleed or other unwanted sounds in between wanted sounds, but Pro Tools, like other DAWs, also makes it easy to cut out or mute clips where no music is playing. There are many ways to do this, starting with the most obvious, which would be to select the unwanted clip and use the Cut command (Command-X on Mac, Control-X on Windows). If you want to remove everything between, say, a series of wood block or clave 'hits', you can use the Separate Clip at Transients command, available from the Edit Menu, which would be a lot faster than individually selecting and deleting the clips between each of these 'hits'! Or you could insert a noise gate plug-in onto the tracks you want to clean up and use dynamics processing instead – which would be the analogue engineer's way of doing things.

> **TIP**
> Sometimes it is better to leave some, or even all, of the incidental noises
> in the recording: it can sound more natural or real that way and this can
> add to the appeal of the music. Recordings made with all the tracks tightly
> gated or with everything cut out of the tracks apart from the featured vocal
> and instrumental phrases can sound too 'processed' for some people – me
> included. It's a bit like the difference between a wholesome homemade
> burger and the MacDuck variety.

Sorting the Tempos

It is possible that a client will bring you a session to mix that has no tempo
set. Although you can just go ahead and mix without any reference to tempo
or bars and beats, it is always a good idea to have a tempo for the music, or a
tempo map if the tempo changes. This way, you can use Grid Mode whenever
you would like to, and you can easily make edits at obvious places such as the
beginning or end of a particular bar or beat. So it is worth spending some time
with the Identify Beat command or using Beat Detective to create a tempo or
tempo map before you start mixing.

Marking the Sections

Markers that say where the verse, chorus, middle, or whatever, sections start
can be invaluable timesavers on any mix session. If there are no markers to
define the various sections of the music or song, then you should create a
suitable set of markers at the outset of your mix session.

Grouping Tracks

You will need to group tracks together for all sorts of reasons during your mix
session. Maybe you want to change the fader levels of all the percussion tracks
at the same time, for example. Or maybe you want to apply a delay effect to
all the backing vocals.

The most basic way of grouping tracks together so that when you move one
fader all the faders in the group move is to use the basic Track Group feature
where you select two or more tracks that you want to group in this way and
press Command-G (Mac) or Control-G (Windows). The next stepup from this is

to create a sub-group by routing the outputs of a group of Audio tracks to the input of an Auxiliary Input channel, so that you can use that channel's fader to control the level of the sub-group. Both of these techniques have their uses and you will find many applications for these during your mixing sessions.

VCA Groups

Experienced mixing engineers will be familiar with the VCA Groups used on analogue mixing consoles to control group or offset the signal levels of other channels on the console. These are available in Pro Tools HD as VCA Master tracks.

Conventional VCA Groups

To understand VCA Groups a little more thoroughly, you need to something about how these were developed. First, VCA is the acronym for a Voltage-Controlled Amplifier. Mixing console designers realized that they could include a set of faders on an analogue console that could be used to send voltages to the amplifiers in selected audio channels so that one of these VCA Master faders would control the audio levels for a group of mixer channels. A group of mixer channels being controlled in this way by a single VCA Master fader is then referred to as a VCA Group. No audio passes directly through the VCA Master fader: this is only used to set the level of the DC control voltage that is sent to control the levels of the faders in the group.

Using just one single VCA Master fader to control a number of mixer channel faders is much more efficient than creating a stereo sub-group using a stereo Aux channel with the individual mixer channel outputs routed to this. A VCA Master is much easier, and cheaper, to implement than an audio channel because it only has one fader that sets a voltage level and some routing and switching circuitry – it has no (relatively) expensive amplifier or signal processing circuitry.

When you lower the VCA Master fader, you are effectively lowering the mixer channel faders in the group that you are controlling. This brings the advantage that any post-fader auxiliary Send feeds are lowered in turn, so the wet-dry balance of any effects that you are applying to these mixer channels stays the same. This is normally what you want to happen and is not the case when you have created a sub-group by routing mixer channel outputs to an Auxiliary Input channel and you use that channel's fader to lower the level of the sub-group.

VCA Master Tracks in Pro Tools

Only available for Pro Tools HD, VCA Master tracks emulate the operation of voltage-controlled amplifier channels on analogue consoles, where a single VCA channel fader would be used to control, group, or offset the signal levels of other channels on the console.

The way this works in Pro Tools is that you group a selection of tracks into a Mix group and then assign this Group to a VCA Master track, as in Figure 7.34. This lets you control the output levels of all the VCA group's member tracks

Figure 7.34
A Group of four handclap tracks controlled by a VCA Master track.

without the need to bus them to an Auxiliary Input track or to the same output path. You can create multiple, nested VCA groups and control the output levels of multiple sub-mixes at the same time – and you can conveniently automate a sub-mix by automating its VCA Master track. A major advantage of using VCA Master tracks in Pro Tools is that they don't use up any mixer DSP resources.

> **NOTE**
> If you wish, you can use a single VCA Master track to control a selection of tracks that are routed to different outputs, which you cannot do by creating sub-groups using Auxiliary Inputs.

Using a VCA Master track has the advantage that changing the VCA fader level will not alter the balance between 'wet' and 'dry' for any tracks using post-fade Aux sends, which would happen otherwise. This happens because when you reduce the fader levels you are reducing the amount of 'dry' signal in your mix – but you are not reducing the send levels to the reverb – so the 'wet' level stays the same.

You will also change the wet/dry balance if you insert a reverb plug-in on an Aux track to which a group of tracks is being bussed and then lower the Auxiliary track's fader – but not if you use a VCA fader to reduce the levels of all the individual faders in this group of tracks while leaving the Auxiliary track's fader untouched (without any gain change).

Because they don't pass audio, VCA Master tracks don't have inputs, outputs, inserts, or sends – they just have a fader along with Solo, Mute, Track Record Enable, and Track-Input Monitor buttons. They also have two pop-up selectors – one for Automation and one for Group Assignment – to let you select which Group of 'slave' tracks to control. Note that on a VCA Master track, the level meters indicate the highest level that has occurred on any of the individual tracks 'slave' it is controlling, not the summed level of all the 'slave' tracks.

Keep in mind that, unlike VCAs on traditional analogue consoles, VCA Master tracks in Pro Tools directly affect their 'slave' tracks, so that the controls on each 'slave' track always show their actual values. In other words, what you see on the 'slave' tracks is what you've got! The levels of the faders show the actual levels as affected by the VCA Master. Even if a particular 'slave' track is a member of more than one VCA-controlled group, the contribution of all the VCA Master faders is summed on the 'slave' track.

Because the 'slave' tracks are grouped together first, using the standard Pro Tools Group command, you would expect that clicking an individual Mute or Solo would Mute or Solo all the Group tracks. However, when controlled by a VCA Master, these 'slave' tracks operate individually, so clicking on just one 'slave' track's Mute or Solo button will only affect that track.

> **NOTE**
> You can over-ride this behaviour and make these controls work as they normally would by de-selecting the 'Standard VCA Logic for Group Attributes' option in the Automation section of the Pro Tools Mixing preferences window.

Also, the solo, mute, record-enable, and Track-Input status on each slave track will correspond with how these controls have been set on the VCA Master. So, for example, when you click Mute on the VCA Master, the individual Mute buttons become engaged on all 'slave' tracks, but in a partially 'greyed out' condition visually, and you will hear no sound. Slave tracks are said to be 'implicitly muted' when you mute the controlling VCA Master, because you are intending, or implying, that you want to mute these slave tracks. If any of the individual slave tracks were muted before you pressed the Mute on the VCA Master, these will remain muted and retain their normal visual appearance. To see how this looks, take a peek at the accompanying screenshot in Figure 7.35 on the next page.

When you click Solo on the VCA Master this button will light up yellow and you will hear all its slave tracks in solo – but the individual slave track Solo buttons will not light up yellow. The way this works is that soloing a VCA Master will implicitly mute all tracks except its slave tracks, thereby indirectly soloing the slave tracks. Also, any explicit solos on the slave tracks will be cleared, leaving them indirectly soloed, and explicitly soloing a slave track while its VCA Master track is soloed will override the VCA Master solo.

> **TIP**
> The way that the Record-enable and Track Input enable buttons work on the VCA Master is as follows: you can toggle Record Enable or input monitor status on and off for any tracks that have been record-enabled using the VCA Master Record Enable or the VCA Master Track-Input button.

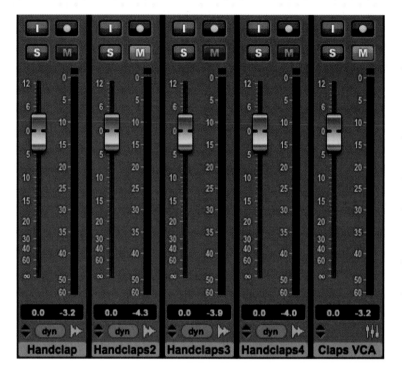

Figure 7.35
Handclaps 2 was muted within the Claps group. Then the VCA was muted. The screenshot shows the appearance of all the mute buttons. The mute button for Handclaps 1, 3, and 4 have a different appearance to indicate that they are being muted only as a consequence of muting the VCA Master and will not be muted when the VCA Master is unmuted.

Using VCAs to Control Other VCAs

It can be very convenient to use VCA Master faders to control two otherwise separate sub-groups or VCA groups from one fader. For example, you might have a sub-group containing a brass section and another containing a string section. You may already be controlling these sub-groups using two separate VCA Master faders, or you may not. Whichever the case, you can assign all the mixer channel faders to one single VCA Master that will let you raise or lower the brass and strings together, as in Figure 7.36.

Another example would be with a choir containing male voices and female voices. You would sub-group the male and female voices separately, set the balances between these, and then use a single VCA Master to raise or lower the level of the whole choir. Yet another example would be to create a sub-group containing drums, then to create a VCA Master to control the drums along with the bass guitar, so that you could balance this rhythm section against the rest of the music.

Figure 7.36
A single Master VCA controls two individual VCA faders that in turn are controlling groups of drums and basses.

Edit Window Mix Automation

The Edit window lets you view automation and controller data either as a Track View to show one or other type of data, or in separate lanes under each track, as in Figure 7.37, in which case you can view as many types of automation and controller data as you wish.

Figure 7.37
Automation and Controller lanes shown below tracks in the Edit window.

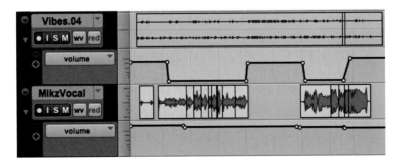

This lets you keep your preferred track view, such as Clip view, visible in the Edit window for each track, ready for last-minute edits, while simultaneously viewing volume automation, for example, in a lane below each track. This can be very convenient during a mixing session, when you might want to make 'tweaks' to the automation graphically onscreen in these Automation and Controller lanes instead of using the Mix window's real-time automation features.

Automation and Controller lanes can be shown or hidden under each track in the Edit window by clicking on the small arrow to the left of the Track controls in Medium or larger views – see Figure 7.38.

Figure 7.38
Show or hide automation lanes using the arrow to the left of the Track controls.

In views smaller than Medium, you can show or hide Automation lanes using the Track options pop-up selector – see Figure 7.39.

On Audio tracks that are in Medium or larger view, you can use the pop-up Lane view selector in each lane to change the type of automation shown in the lane – see Figure 7.40.

Figure 7.39
A Track Options pop-up is located at the top left of each Track's controls area in the Edit window.

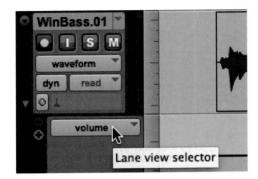

Figure 7.40
Lane view selector.

Choices include volume, volume trim, mute, and pan, along with automation controls for any sends or plug-ins that are active – see Figure 7.41.

Figure 7.41
Lane view selector pop-up for an Audio track.

In the example given in Figure 7.42, an Instrument track is shown with a Controller lane open for MIDI Velocity and an Automation lane open for Audio Volume.

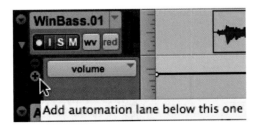

Figure 7.42
An Instrument track with Controller lanes for MIDI Velocity and Audio Volume. You can click on the + button to add another automation lane.

If you want to add more lanes below a track, you can click on the + button in an existing lane. To remove a lane click on the minus (−) button. The Lane view selector for an Instrument track lets you select from the various MIDI Controllers and Audio Automation types – see Figure 7.43. The Lane view selector for a MIDI track just lets you select from the various MIDI Controllers.

To reorder lanes on-screen, drag the lane control sections to new positions in the Edit window. You can resize the heights of all the Automation and Controller lanes for a track by clicking and dragging the bottom line of any Lane Controls column up or down. To resize the height of a single Automation or Controller lane for a track, hold down the Control (Mac) or Start key (Windows) while you select the Lane Height setting. To increase or decrease the height of any lane that contains the Edit cursor or an Edit selection, hold down the Control (Mac) or Start key (Windows) and click the Up- or Down-Arrow key.

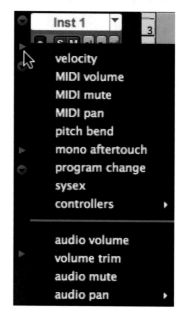

Figure 7.43
Lane view
pop-up selector for
an Instrument track
shows the MIDI
Controllers and Audio
Automation types
that can be selected.

Clip Gain

Pro Tools has a very useful feature that lets you quickly adjust the gain (the amount of boost or cut) of the audio signal level in any Clip in the Edit window. This gain adjustment operates pre-fader and before any plug-in processing and the clip gain settings stay with the clip, which means you can move, and copy and paste clips with their corresponding clip gain settings. Clip gain is set to 0 dB by default and can be adjusted from −144 dB to +36.0 dB.

You can always adjust the gain of a clip using track automation, but it is often much simpler to quickly adjust the gain on individual clips to match these up – perhaps when you have comped a vocal or a guitar solo from several different takes where the levels may have been different from take-to-take – as in Figure 7.44.

You can clear the clip gain settings for the current Edit selection, resetting the clip gain to 0 dB, by pressing Control-Shift-B (Mac) or Start-Shift-B (Windows),

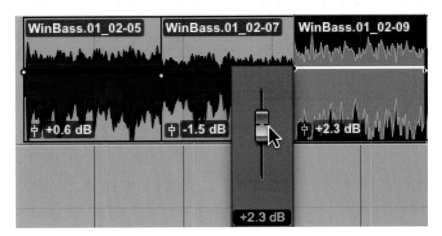

Figure 7.44
Three clips with different static clip gain settings.

or by holding down the Option (Mac) or Alt (Windows) key and clicking on the clip's Clip Gain Fader icon. For clips only partially included in the Edit selection, only the clip gain settings within the Edit selection are affected.

It is also possible to arrange for the clip gain to be automatically (dynamically) adjusted as the clip plays back instead of simply applying a static clip gain setting that does not change while the clip plays back.

Clip Gain Info view

When the Clip Gain Info view is enabled from the Clip sub-menu in the View menu, the Clip Gain Fader icon is shown at the beginning of the clip, in the lower left corner. If the clip uses static clip gain, the static Clip Gain value (−144 dB to +36.0 dB) for the clip is displayed to the right of the Clip Gain Fader icon as in Figure 7.45.

Figure 7.45
Clip Gain Icon.

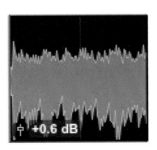

For a clip that has dynamic clip gain (using breakpoint gain settings), the Clip Gain value is not shown. Clip Gain commands (including the Clip Gain Fader and its Right-click commands) only apply to the clip for which you are adjusting the settings.

You can show or hide Clip Gain using the Clip sub-menu in the View menu or by pressing Control-Shift-'=' (Equal) (Mac) or Start-Shift-'=' (Equal) (Windows).

Clip Gain Fader Icon Right-click Menu

If you Right-click the Clip Gain Fader icon for any individual clip, a pop-up menu appears (see Figure 7.46) that allow you to bypass or clear the clip gain settings; or to Render the settings (creating a new clip that replaces this clip in the process); or to show/hide the Clip Gain Line.

The reason that Pro Tools lets you bypass the current clip gain settings for any selected whole clip is so that you can hear the clip without any clip gain adjustments (and most importantly) without losing your current clip gain settings. There will be occasions where you will find his command very useful while making decisions about the correct level for the clip.

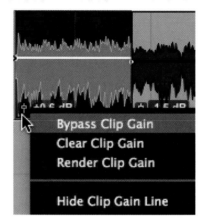

Figure 7.46
Clip Gain right-click pop-up menu.

> **TIP**
> Be careful to point exactly at the Clip Gain Fader icon or the much larger right-click pop-up menu for the Clip will appear. You can also access the Clip Gain sub-menu from this menu, but it is faster to use the Clip Gain Fader pop-up menu.

Once you have made your decisions about the optimum gain settings, you can 'render' the current clip gain settings for any selected whole clip. This 'rendering' process creates a new clip, applying the current clip gain settings to the clip during this process, then sets the clip gain settings for the new clip to 0 dB. If you make an Edit selection that encompasses a number of whole clips, these will all be rendered accordingly.

> **NOTE**
> When rendering clip gain settings, the resulting clip provides handles for trimming out the clip based on the Default Handle Length settings in the Processing Preferences.

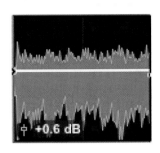

Figure 7.47
Clip Gain Line.

Clip Gain Line

The Clip Gain Line (see Figure 7.47) lets you edit the clip gain settings for any clip using breakpoint editing, in a similar way to using track-based volume automation. However, unlike track-based volume automation, the clip gain settings are always associated with the clip rather than with the track. To show or hide the Clip Gain Line, press Control-Shift-'-' (Hyphen) (Mac) or Start-Shift-'-' (Hyphen) (Windows).

> **NOTE**
> The Clip Gain Line is not shown when the track height is set to Mini or Micro.

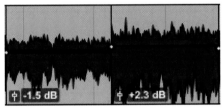

Figure 7.48
Clip Gain Lines in adjacent clips before crossfading.

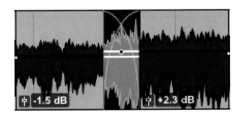

Figure 7.49
Clip Gain Lines in adjacent clips after crossfading.

Clip Gain Line and Crossfades

When cross-fading between clips, the Clip Gain Line for the first clip carries through the fade-out segment of the crossfade and the Clip Gain Line for the second clip carries through the fade-in segment of the crossfade. Consequently, crossfades can display two Clip Gain Lines, one for the first clip and one for the second clip – see Figures 7.48 and 7.49.

Editing Clip Gain

Pro Tools lets you adjust the gain settings for a clip using the Clip Gain Fader for quick and easy adjustments, or using breakpoint editing on the Clip Gain Line for detailed clip gain control. Just drag a breakpoint up or down to change the gain and to left or right to change the position in time and hold down the Command (Mac) or Control (Windows) key while adjusting the Clip Gain Fader for fine control.

To add a Clip Gain breakpoint at the current Edit location press, press Control-Shift-E (Mac) or Start-Shift-E (Windows).

To delete a clip gain breakpoint using the Grabber or Pencil tools, Option-click (Mac) or Alt-click (Windows) on the breakpoint you want to delete.

Figure 7.50
Clip Gain Line with multiple breakpoints.

Using the Grabber tool, you can add a clip gain breakpoint by clicking at any point on the Clip Gain Line – see Figure 7.50. You can also use the Pencil tool to create new breakpoints by clicking anywhere on the Clip Gain Line, and with this tool, you can also use the Free Hand, Line, Triangle, Square, and Random Pencil Tool shapes for drawing clip gain.

After you have inserted multiple clip points onto a clip, you can raise or lower the gain of all these by the same amount using the Clip Gain Fader – see Figure 7.51. Hold down the Command (Mac) or Control (Windows) key if you need fine control.

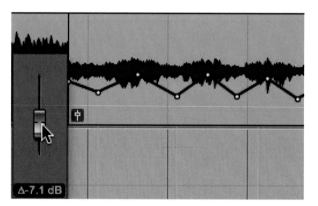

Figure 7.51
Clip Gain Line with multiple breakpoints being adjusted simultaneously using the Clip Gain Fader.

> **NOTE**
> You can always use the Trim tools to adjust all selected breakpoints up or down by dragging anywhere within an Edit selection. Unlike track-based volume automation, which scales when trimming, clip gain provides true trimming (where clip gain settings maintain their fixed relations to one another when trimming).

Fine-tuning Clip Gain

Pro Tools lets you nudge the selected clip gain settings up or down by the Nudge Clip Gain By amount that you can specify in the Pro Tools Editing Preferences. You can also nudge the selected clip gain settings back or forward in the clip by the specified nudge amount.

To nudge the selected clip gain up, hold down the Control and Shift (Mac) or Start and Shift (Windows) keys and press the Up Arrow. To nudge the selected clip gain down, hold down these modifier keys and press the Down-Arrow instead. If you have a mouse with a scroll wheel, hold down these modifier keys and you can use the scroll wheel to nudge the selected clip gain settings up or down.

To nudge the selected clip gain back, hold down the Control and Shift (Mac) or Start and Shift (Windows) keys and press either the '–' (Minus) or the M key, or the ',' (Comma) key. To nudge forward, hold down these modifier keys and press either the '+' (Plus) or the '.' (Period) key, or the '/' (Slash) key.

Pro Tools HD Clip Gain Features

Pro Tools HD lets you convert (or coalesce) clip gain settings to track-based volume automation, as well as letting you convert (or coalesce) clip track-based volume automation to clip gain settings.

To convert (or coalesce) clip gain settings to track-based volume automation, select a clip or make an Edit selection then choose Convert (or Coalesce) Clip Gain to Volume Automation from the Edit menu's Automation sub-menu – see Figure 7.52. To convert (or coalesce) track-based volume automation to clip gain settings choose Convert (or Coalesce) Volume to Clip Gain instead.

Any clip gain settings above +12 dB are lost when converted (or when coalesced) to volume automation. Also, clip gain settings within crossfades are crossfaded as part of the volume automation.

Figure 7.52
Clip Gain commands
in the Edit menu's
Automation
sub-menu.

When converting (or coalescing) clip gain settings to volume automation, the clip gain settings are cut from the clip and pasted to (or coalesced with) track-based volume automation at the same timeline locations as the clip.

When converting (or coalescing) track-based volume automation to clip gain, the volume automation is cut from the track volume automation playlist and pasted to the clip (or coalesced with the selected clip gain settings). Once volume automation has been converted (or coalesced) to clip gain, the clip gain settings stay with the clip when moved, cut, copied, or pasted (and, if coalesced, the volume automation is set to 0 dB for the selection).

Mix Moves

Tweaking the Mix

Having completed all your preparations, the final balancing session often becomes relatively easy. In the early stages of your mixing session, you will have chosen most of the elements you want to work with and you will have set up the effects you want to use. It can take some time to set up sends to external effects units and Auxiliary tracks with chains of plug-ins to process the drums, the vocals, and the various instruments – especially if you are looking for that 'special' reverb sound or combination of delays, or if you are trying to find the right EQ settings and pan positions to make particular instruments 'sit' properly in your mix, or stand out as features. With a typical pop song, you should allow at least half a day to get everything more or less in place. Then you might take a break for half an hour or so before coming back to do the final balancing of levels and tweaking of effects. Here you should be concentrating on the most important tracks – typically the lead vocal and any featured solo instruments.

Make sure that all the tracks you are not using are either removed from the session or made inactive and hidden and set the rest to the minimum track height – just leaving the lead vocal, backing vocal, and perhaps some solo instruments at medium track heights. It is also a good idea to display the Volume Automation curves in the chosen tracks so that you can see what is going on and so that you can manually adjust these volumes at any stage during the mix session, as in Figure 7.53. Alternatively, you can use the Automation lanes to display volume or other automation or controller data in the Edit window.

If you are using the real-time automation features, you will start out in Auto Write mode. You probably won't get all your moves correct during the first pass,

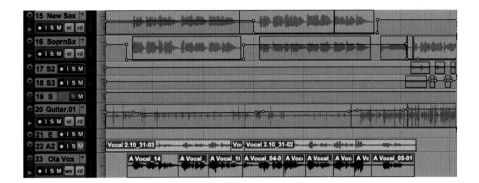

Figure 7.53
Pro Tools Edit window during a mixing session with the volume automation curves for the Sax and Guitar visible – ready for the mix engineer to tweak.

so you can use Auto Touch mode to refine the sections that need changing. As you get closer to what you are looking for, you might use the Trim features to fine-tune the settings even further. For example, if you like the way the guitar rhythm gets softer in the verse and louder in the choruses, but you decide that overall the guitar is too loud, you can use Trim mode to bring the overall level down on the guitar track while retaining the relative automation 'moves'.

I usually end up going into the Edit window to manually edit the automation breakpoints to achieve exactly the right result. Here, it is so easy to see exactly where the vocal and instrumental phrases lie and to edit the automation curves to do exactly what you want them to do.

TIP
To move all the automation breakpoints in a particular track or section of a track up or down while preserving relative levels, choose the Selector tool first and drag the mouse to highlight the range of interest. Then choose the Trimmer tool, point the mouse anywhere along the line of breakpoints and drag the whole line upwards or downwards as in Figure 7.54.

Figure 7.54
Manually trimming a selection of Automation breakpoints.

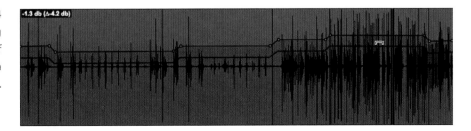

Finalizing the Mix

Once you have your mix sounding the way you want it, it's time to create a stereo 'master' mix. You have various options here. You can 'bounce' to disk, or you can record from the main stereo outputs of your Pro Tools interface into a stereo mastering recorder such as a CD-recorder or a DAT recorder, or you could even record to a 1/2-inch or 1/4-inch analogue tape recorder.

> **TIP**
>
> Recording your mix directly into Pro Tools or to an external recorder allows you to adjust the mixer controls in real time and to include audio coming in from external sources.

Quite often, you will want to create a full mix with everything in, plus a mix minus the lead vocal, plus a mix minus the lead vocal and the backing vocals. This allows a singer or a vocal group to do personal appearances singing to this 'backing track', with or without the backing vocals.

Some producers like to create so-called 'stem' mixes. This is a common way of working in post-production and film work, where dialogue, sound effects, and music are typically kept separate until the final 'dub' where the film soundtrack is mixed. This allows easy replacement of the dialogue with foreign languages and also allows the balance between the sound effects, dialogue, and music to be adjusted for film releases in different parts of the world where the preferred balances between these are different.

It can also make sense to create stem mixes for music productions. For example, you could make a sub-mix of all the backing vocals, of the strings, of the brass, of the rhythm section, or of whatever best suits your purpose. Typically, you would also make a stereo mix of the lead vocals with all the effects, panning and level changes applied. With all these stem mixes available, you can easily make new mixes with the lead vocals louder or softer, with the backing vocals louder or softer, and so forth.

In recent times, there has even been a trend toward using stem mixes at mastering sessions so that the mastering engineer can be involved in the final balance decisions, although only a few mastering studios offer this kind of session.

Yet another popular technique is to route the entire mix into an Auxiliary Input, using this as a 'sub-master' before routing this in turn to the

Master Fader. The advantage here is that you can use pre-fader Inserts on the Auxiliary Input to add dynamics processing to the whole mix. Remember that Inserts on a Master Fader only work in post-fader mode, so if you insert a dynamics processor, adjust this to your liking, then make any subsequent changes to the level of the Master Fader (which will happen on a fadeout), the compressor's threshold setting won't take these changes into account.

Recording Mixes to Tracks Using a Sub-master

You can record any mix or sub-mix directly into your Pro Tools session by routing the mix or sub-mix to available audio tracks in the same session. This method allows you to add live input to the mix as well as adjust volume, pan, mute, and other controls during the recording process.

Figure 7.55
Output path pop-up
selector.

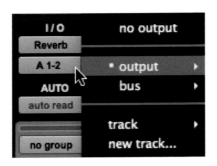

The best way to set your session up if you intend to record your mix to Pro Tools tracks is to route all the tracks you want to include in your mix to an Auxiliary Input track first. The fastest way to do this is to select all the tracks you wish to include in your mix then hold the Option and Shift (Mac) or Alt and Shift (Windows) keys and click on any track's Output Path Selector – see Figure 7.55.

When the pop-up menu appears, choose 'New Track…' and the New Track dialogue will appear with a Stereo Aux Input track already selected for you. Type in a suitable name, such as Sub-Master (see Figure 7.56), and OK this to create a new stereo Aux track with all your chosen tracks routed to this.

The output from this 'sub-master' track will automatically be routed to your main outputs, so that you can monitor your mix, so you will also need to route an additional output from the sub-master to the inputs of the tracks that you will record your mix to.

Figure 7.56
New track dialogue
set up to create a
Stereo Aux Input.

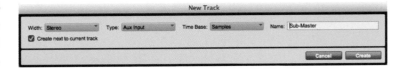

Go to the I/O setup window and rename an available bus pair to use as your Bus Record route. Now add a second output to your sub-master track by Control-clicking (Mac) or Start-clicking (Windows) on the Audio Output Path Selector and choosing Bus Record from the 'bus' sub-menu. Check that a (+) symbol has appeared on the Output Path Selector to indicate that these multiple destinations have been selected.

Remember that recording a sub-mix to new tracks requires an available voice for each track that you want to record, so you will need to make sure that you have enough voices available to play back all the tracks that you want to record and enough additional voices available to record your mix tracks. And if you are using a Pro Tools HDX system, keep in mind that Native plug-ins placed on Auxiliary Input, Master Fader, and Instrument tracks will also use additional voices and may introduce extra latency based on the H/W Buffer Size.

Create a stereo audio track to record to, setting the pans to hard left and right and set the input to the bus path from which you are recording (named Bus Record in this example) and set the output to your main output monitoring path (Output pair 1–2 in this example).

> **TIP**
> Make sure that Loop Record is not enabled and that 'Link Timeline and Edit Selection' is selected from the Options menu.

When you have the tracks set up and routed, select the audio to record, keeping in mind that the start, end, and length of the recording can be based on the cursor location or Timeline and Edit selections. Very conveniently, an Edit selection-based recording will automatically punch in and out of recording at the selection start and end. Be sure to include time at the end of your selection for reverb tails, delays, and other effects to die away. If you don't make an Edit selection, recording will begin from the location of the playback cursor and will continue until you press Stop.

With this setup, because you are monitoring the output of the sub-master through the main outputs, you will need to make sure that you are not simultaneously monitoring the outputs from the tracks you are recording onto, so you should mute these – see the accompanying screenshot, Figure 7.57.

Figure 7.57
Bus recording mixer tracks via an Auxiliary Input track to a stereo audio track.

Trimming the Bus Feeding the Sub-master and/or Master Fader

If you see the clip indicators showing red on the Master Fader or showing yellow on the sub-master if you are using one, you can use another Master Fader to trim the level of the bus before it reaches the sub-master or Master Fader. Alternatively, you could manually adjust the levels of the individual tracks being summed onto the bus, or you could use a VCA Master Fader to trim the levels of all the source tracks that are being summed onto the bus.

So which to choose? If you are only working with a small number of tracks, manually adjusting the individual channels is a practical proposition, but with larger mixes this can be very time-consuming. Using a VCA Master to adjust the levels of the source tracks feeding the bus can be a good option, especially when you are creating sub-mixes of tracks that are already grouped.

Using Master Faders also allows you to meter the bus level passing through this fader, so I prefer this option in most situations.

Track Output Window and Path View Meter

When you are working on your mix, setting up sub-mixes using Aux tracks as sub-masters or to make stem mixes, it can be helpful to open the Output Window for the Aux track so that you can keep an eye on the meters to watch for clipping at the input. Use Pre-Fader Metering mode for this, which you can select from the Options menu. If you expand the track's Output window to view the Path View meter as well, then you can see if you are also clipping the output path.

To open the track's Output Window, click on the Output Window – see Figure 7.58. Button (the tiny fader icon at the right side of every Audio Output Path Selector). This is a 'floating window' that you can position anywhere that is convenient on your screen while you are moving things around in the Mix window.

Figure 7.58
Opening the Track
Output window.

NOTE

Output windows provide the essential track mixing controls (track fader, pan, automation, solo, and mute controls), as an alternative to Mix and Edit window views. These are useful in large sessions to leave important tracks in an anchored location, unaffected by Mix and Edit window (or control surface) banking. To open multiple Track Output windows, disable the red target button on the first and subsequent windows that you open. When the pan controls are in the linked state, all sides match changes to any other side's Pan control. Unlinked mode provides discrete adjustment of individual sides or channels. When linked and set to Inverse mode, panning moves are inverted or reversed in the other channels. Inverse panning reflects one side's location and direction in the other side, so if you pan one side of a stereo track output from right to left, the other side will exactly mirror that movement and pan left to right. Output windows provide standard selectors for path, automation, and other controls in the top Editor area of the Output window. The Track Selector provides access to any audio track, Auxiliary Input, or Master Fader in the session. The Output View Selector provides access to other outputs in the track, if any, displaying the selected output in the current window. The Output Path Selector allows you to assign the output path for the current track. Automation Safe protects level, pan, and other controls from automation overwrites by placing them in Automation Safe mode.

Figure 7.59
Path Meter View button.

If you then click on the small green zoom button at the top of the window, this will open the track's Path Meter View as well – see Figure 7.59.

Here you will see meters (as in Figure 7.60) that show the level of the summing bus to which the track's output is assigned – either an Aux Input bus or an audio interface output. This meter is the same meter that you can see on the Master Fader that controls this bus or output path and shows the level produced by the sum of all the tracks routed to that bus or output path.

You can use this meter to see if you are clipping any bus or output path, and it can be very helpful to keep this window open and in view when the Master Faders controlling these bus or output paths are not in view in the Mix window.

Figure 7.60
Output window expanded to show the Path View meters belonging to the summing bus to which the track's output is assigned – in this example, showing the meters for the main output bus pair, A1-2.

> **NOTE**
> The green Path Meter View button at the top of the window expands the window to open the Path Meter View. The meters displayed here show the levels of the selected path – not the levels of the track. If you have multiple output paths available in your session, you can choose which output path to view here using the pop-up selector underneath the meters.

Exporting Bus-recorded Clips

After you have recorded your stems or mixes to audio tracks in your session, you may find it convenient to export these audio clips using the 'Export Clips as Files' command which can be accessed by Right-clicking any clip in the Clips list – see Figure 7.61.

Figure 7.61 'Export Clips as Files' command.

The Export Selected dialog (see Figure 7.62) that appears when you choose this command allows you to convert clips to a different audio format, sample rate, or bit depth and to export left and right audio files as multiple mono or as stereo interleaved files.

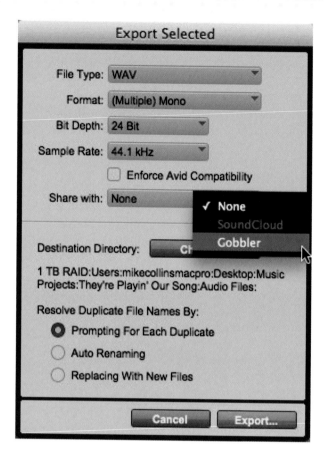

Figure 7.62
Export Selected
dialogue.

> **NOTE**
>
> Pro Tools HD also let you bounce multi-channel interleaved files of any supported file type.

When you export clips to a lower bit depth, Dither (with or without Noise Shaping) is applied, using the Avid Dither plug-in by default. You can also 'share' files with Gobbler or SoundCloud.

Bouncing to Disk

Bounce to Disk Command

You can use the Bounce to Disk command in Pro Tools to directly create a stereo file on disk. To do this, you mute everything but the tracks you want to bounce, make sure that all the levels, pans, and any effects and automation are the way you want them to be, assign the outputs from all the tracks to the same pair of outputs, then select the Bounce command from the File menu and bounce your mix from this output pair to disk.

Although you can hear the bounce as it is being created, you can't adjust any controls during a 'Bounce to Disk'. So you should only use this method if you don't need to interact with any mixer controls during the bounce. Also, you cannot include audio coming in from external inputs during a bounce to disk, so if you need to do this you will need to record to new audio tracks.

> **NOTE**
> If you have Low Latency Monitoring enabled, only your audio tracks are included with the Bounce to Disk command – any Auxiliary Input and Instrument tracks in your session will be ignored. To include Auxiliary Input and Instrument tracks, disable Low Latency Monitoring before using Bounce to Disk.

New Bounce Features

Several new features have been added to the Bounce dialog in Pro Tools 11. For example, you can name the file you are about to create directly in the Bounce dialog, bounce multiple output and bus paths simultaneously, bounce from sends and track outputs, bounce multiple sources to QuickTime – and you can automatically share (send copies of) your bounces directly to SoundCloud or Gobbler.

The new Add MP3 option in the Bounce to Disk dialog lets you simultaneously create an MP3 version of each bounce source in addition to the selected File Type – see Figure 7.63 – so now you can make a WAV file and an MP3 at the same time!

One thing to watch out for here is that when you use the Add MP3 option and you are adding the bounce to your iTunes Library or sharing with Gobbler or SoundCloud, the selected File Type (WAV, etc.) is used rather than the MP3. So if you want to share the MP3 file, you do have to select MP3 as the File Type.

To do this, make a Timeline selection to define the range to be bounced, choose Bounce to Disk from the File menu, configure the various options, and click Bounce. You will be presented with the MP3 dialog – see Figure 7.64 – where

Figure 7.63
Add MP3 option.

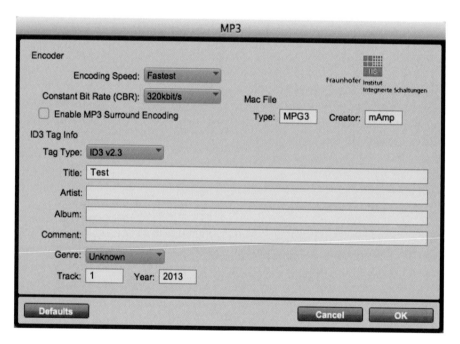

Figure 7.64
MP3 Dialog.

you can choose the type of encoding and enter various metadata. When you OK this dialog, two files are created and placed inside a 'Bounced Files' folder inside your session folder.

Bouncing to Disk

The standard Bounce to Disk feature in Pro Tools 11 lets you bounce to Broadcast WAV, AIFF, MP3, or MXF (Material Exchange Format) file types. You can bounce in real-time so that you can hear your mix playing back, which allows you to check this for any problems during the bounce, or you can do

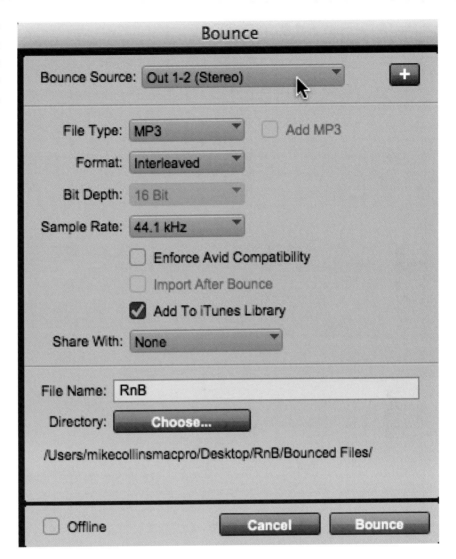

Figure 7.70
Bounce Dialog
showing the
Bounce Source
pre-configured.

Bounce Destinations

The default destination for bounced files is the Bounced Files folder located within your project's Session folder, but you can always choose a specific directory by clicking the Choose… button and navigating to a directory of your choice.

Pro Tools also lets you add your mix to your iTunes library using the Add to iTunes Library option in the Bounce to Disk dialog. When this option is selected, the bounced file is first saved to the chosen bounce destination, then it is automatically imported into iTunes the next time you launch the iTunes application. You can specify the location of your iTunes library using the new iTunes Library Folder setting in the Pro Tools Operation preferences.

NOTE

The Add to iTunes Library option is only available if the Bounce Source is set to a mono or stereo path, and the Format is set to Mono (Summed) or Interleaved. Also, if iTunes is not installed on your computer, this option does not do anything.

Bounced files can also be 'shared' with, or to put this in ordinary language, they can be simultaneously sent to one of two additional destinations: SoundCloud or Gobbler.

SoundCloud is a 'cloud-based', 'social audio sharing' online audio distribution platform, established in Berlin, Germany, since 2007. 'Cloud-based' means that the files you send to SoundCloud are stored on remote servers from which they can subsequently be streamed for real time playback or downloaded. 'Social audio sharing' means that it has features for sharing files with social media such as Facebook, Tumblr, Google +, Pinterest, and Twitter. SoundCloud sound files can also be embedded just about anywhere and individual files or sets of files can be distributed to individuals or groups, or to the entire SoundCloud community, which is already very large and continually growing. For example, on January 23, 2012, SoundCloud announced on their blog that they had 10 million registered users. Apps are also available for iPhone, iPad, and Android platforms.

Pro Tools lets you share your mix with SoundCloud using the Send To SoundCloud option in the Bounce to Disk dialog or in the Export Selected dialog. When this option is selected, the bounced (or exported) file is automatically uploaded to your SoundCloud account. If you do not have an account, you can create one for free.

Having a thorough understanding of how the meters and faders work, and how you can use Master Faders to prevent clipping on outputs and buses is probably the most important section in this chapter. After all, mixing is very much about getting the levels of the mix elements balanced to your satisfaction – without the digital waveform clipping that so often produces such harsh sounding mixes on commercially distributed music today. So I will leave you with these thoughts…

In this chapter

CHAPTER 8

Pro Tools 11 new features

Moving to Pro Tools 11

When moving to Pro Tools 11 from previous versions, you will immediately notice how much better the performance is when it comes to running large sessions or using lots of virtual instruments with its new Avid Audio Engine. The new Avid Video Engine – as used with Avid's Media Composer systems – also makes workflows much more efficient for those working to picture. And now that Pro Tools 11 is a 64-bit software application, the previous 4GB RAM limitation with the 32-bit versions of Pro Tools no longer applies – you can use as much of the RAM you have installed on your computer as and when you need this – so you can have more tracks and more instances of virtual instruments now that additional memory can be addressed. Pro Tools 11 also makes much more efficient use of multiprocessor cores and includes dynamic processing (described later), so it can handle much more complex MIDI sessions with lots more virtual instruments.

Pro Tools 11 now time-stamps all automation, making sure that each playback is as accurate as possible, and Pro Tools 11 can now write automation while recording – a great benefit for post-production engineers and for music tracking mixers who want to control groups with VCA Masters while recording.

Generally speaking, you will also find that working with Pro Tools is much faster and simpler as a consequence of the many updates to the user interface and key commands. There are many small but extremely useful improvements. For example, previous versions of Pro Tools only supported up to 32 levels of undo, but now you can specify the actual number of undo's in the Editing preferences up to a maximum of 64.

You can now bypass the Sends while you are setting things up, either individually or collectively, and you can meter any dynamics processing that you are using on a track – and you can do this either individually or collectively when using more than one dynamics processor. The new Workspace has also been drastically reorganized to speed up the workflow.

Avid Audio Engine

The Engine

For Pro Tools 11, Avid has developed a new Audio Engine – which is also used for Avid's Media Composer. This provides much greater compatibility and interoperability between all of Avid's audio and video systems. According to Avid, the new Audio Engine delivers much more processing power for using virtual instruments and plug-in effects than the Digidesign Audio Engine (DAE), which was used with previous Pro Tools systems. The Avid Audio Engine also provides ultra-low latency record monitoring. This uses brand-new code that can take advantage of all the processors' cores and evenly distribute the load over them. You no longer have to choose how many processors to use in the Playback Engine dialog as the engine can handle all of them efficiently.

Configuring the Audio Engine

You can configure the Avid Audio Engine in the Pro Tools Playback Engine dialog, which you can access from the Setup menu. Here you can choose the correct Playback Engine for the audio hardware you wish to use for playback. This could be a Pro Tools HDX card, any third-party hardware – or the built-in audio in your Mac or PC. A pop-up selector lets you choose a Hardware Buffer Size appropriate for your session.

Low-Latency Input Buffer

Pro Tools 11 provides a new low-latency input buffer that ensures ultra-low latency record monitoring – avoiding the need to constantly adjust the buffer or the active plug-ins and tracks in the session. This feature is not user-configurable. Instead, with Pro Tools HD, the low-latency input buffer is automatically engaged for record-enabled and input monitor–enabled tracks.

The way this works is that all of the playback tracks use a fixed (relatively high-latency) playback buffer. Any tracks that have a live input, whether an audio track with a physical input assigned in record or track input mode, or a virtual instrument track in record, are automatically switched to use the low-latency buffer. The size of this buffer can be adjusted using the hardware (H/W) buffer size pop-up selector in the playback Engine dialog – see Figure 8.1.

According to Avid Solutions Specialist Simon Sherbourne, 'Here's how the buffer thing works: Pro Tools used to have a single native processing buffer, so setting it was always a compromise between latency and processing.

Figure 8.1
Playback Engine
Dialog.

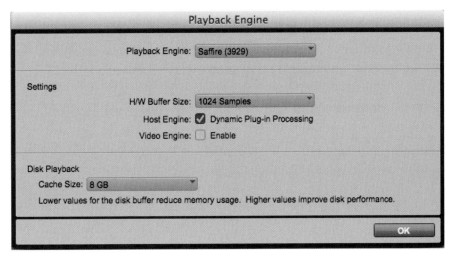

Now there are two co-existent buffers (or "domains") that Pro Tools can assign tracks and native plug-ins to on the fly. One is fixed and high, the other is user-definable. Pro Tools only assigns plug-ins and tracks to the lower buffer where necessary, so you can typically leave the user buffer low without compromising performance.

Whichever virtual Instrument/MIDI track is currently the Thru track (usually whichever is selected if you have that preference on) will also switch to the low latency domain. "Input" buffer is not really how we describe it, as it's actually the host-processing buffer. It affects several things: native plug-in latency; MIDI-to-audio latency when playing virtual instruments; audio throughout latency on input monitoring/recording tracks in Native systems; and the latency in an HDX track where you move between DSP and host processing. The buffer does not apply to input/rec tracks on HDX systems unless there is a native processing plug-in on the track.

It's really clever, as the whole signal path "downstream" of the live input will also switch to the low buffer: i.e. plug-ins on any tracks to which the input monitoring track is routed (directly or via a send) will switch to the low latency domain'.

Delay Compensation

If you are upgrading from a previous version, you may be wondering where to find the Delay Compensation settings. With Pro Tools 11, Automatic Delay Compensation is always enabled with the maximum setting – currently

16,384 samples. Pro Tools 11 also provides Automatic Delay Compensation on sends, which was not the case with previous versions. With Pro Tools HD 11 software, and if you are using Pro Tools HDX hardware, the software will automatically compensate for signal delays in side-chain processing.

Dynamic Plug-in Processing

In the Playback Engine dialog, there is an option to enable Dynamic Plug-in Processing on the Host Engine, and if you are using Pro Tools HDX hardware, this also lets you select the Number of Voices.

The way Dynamic Plug-in Processing works is that if you are using plug-ins on a track, but there are sections on the timeline where there is no audio in the track, Pro Tools will only use the CPU to apply this plug-in processing to the sections of the track where the audio is present. This prevents the CPU from processing nothing and frees up this processing power for use elsewhere.

Avid Video Engine

If you plan to work with video, you will need to enable the Avid Video Engine in the Playback Engine Dialog.

Pro Tools 11 supports the Avid Video Engine – the new video engine for both Pro Tools and Media Composer. As with the new Avid Audio Engine, this enables greater compatibility and interoperability across the full product line of Avid audio and video editing solutions.

With the Avid Video Engine, Pro Tools 11 provides support for a wide range of HD video formats on Pro Tools video tracks without transcoding (including Avid DNxHD). Pro Tools 11 also provides support for monitoring DNxHD and QuickTime media using Avid Nitris DX, Mojo DX, and other Avid qualified third-party video interfaces such as AJA and Black Magic Design.

A boon for post-production engineers putting sound to picture, this makes it possible to immediately import HD video in these industry-standard formats without transcoding and to watch in either full-screen or low-resolution quality directly from the timeline. This means you can use a low-resolution version of the video for much of your work but then switch momentarily to the high-resolution video to check lip sync and important automation moves or edits to make absolutely sure that these are 'spot on'!

Avid Satellite Link

Avid's Satellite Link software is included for free with Pro Tools HD 11. Satellite Link lets you link up to 12 Pro Tools systems over an Ethernet network, so that you can cue, play, and stop the transports, make play selections, and solo tracks across any of the systems from any linked workstation. Or, you can link up to 11 Pro Tools systems and an Avid Video Satellite system (Avid Media Composer or Avid Symphony Nitris DX) or using a Pro Tools Video Satellite system.

New Bounce Features

Bounce to Disk

Several new features have been added to the Bounce dialog in Pro Tools 11. For example, you can name the file you are about to create directly in the Bounce dialog, bounce multiple output and bus paths simultaneously, bounce from sends and track outputs, bounce multiple sources to QuickTime – and you can automatically share (send copies of) your bounces directly to SoundCloud or Gobbler. The new Add MP3 option in the Bounce to Disk dialog lets you simultaneously create an MP3 version of each bounce source in addition to the selected File Type – so now you can make a WAV file and an MP3 at the same time!

Offline Bounce

Pro Tools 11 provides an Offline (i.e. non-real time) Bounce to Disk option that lets you Bounce to Disk faster than real time in most cases – depending on the nature of the session. Pro Tools HD 11 users need to be aware that AAX DSP plug-ins will not work with the offline bounce feature, although AAX Native plug-ins used in a Pro Tools HD 11 session will.

Advanced Metering

New Design

One of the most welcome improvements for Pro Tools 11 is the new meter designs for audio, Auxiliary Input, Instrument, and Master Fader tracks.

The first thing you will notice about the new meters is that they are about 30% taller than previously – giving the user more detail about the level of a channel. The channel faders in the mix window are about 30% taller as well. Also, the graphics resolution is higher (there are more dots per inch) for the meters – and the rest of the Pro Tools user interface has also been upgraded to match.

Meter Types

There are four basic types of meter: Sample Peak, Pro Tools Classic, Venue Peak, and Venue RMS. With Pro Tools HD, you also get Linear and Linear Extended, RMS, VU and Digital VU, three Bob Katz K-scale meters, several PPM meter types, and PPM Digital (taken from the Avid Euphonix System 5 console).

Meter Scales

Meter scales can be changed on the fly while the transport is running by right clicking on a meter in the mix window and selecting a new meter type. The Master meters can be set to a different scale from the Channel meters in a session. Users can also add an output meter to the Edit window toolbar and to the Transport window. The ability to see volume and headroom at the same time on the Mix Page. Clip indication has changed as well, with the yellow clip indicator showing internal mixer levels over 0 dB while a red clip shows that an output or the converter is clipping. HD software also lets you show a gain reduction meter for every track. There is also a new option for HD to see gain reduction in the Insert assignment view so you can see gain reduction activity on a plug-in by plug-in basis.

Workflow Enhancements

Mute Sends Commands

Pro Tools 11 provides new commands in the Track menu to let you mute all the sends in your session, or just sends A–E or F–J.

These are also available as key commands in the mix window: Press Shift + Q to mute all sends; press Shift + 3 to mute sends A–E; press Shift + 4 to mute sends F–J.

It is also possible to mute a group of sends on a particular track using a key command. If you Control-click (Mac) or Start-click (Windows) a track Send

Assignment button in the Mix or Edit window, this will not only Mute the send you have clicked on but will also mute all the sends in positions below it **on this** track.

If you hold the Option (Alt) button while using this command, a single click on any send will mute all the sends in your session. So, if you Control-Option-click (Mac) or Start + Alt-click (Windows) on a Send Assignment button in the Mix or Edit window, this will not only Mute this send, it will also mute all sends below it **on all** tracks. In other words, this lets you mute all the sends in your session with a single click.

> **NOTE**
> To un-mute the sends, simply apply the Mute commands a second time.

Bypass Inserts Commands

When you are working on a mix, it can be very convenient to bypass various inserts while you are setting up your mix session to help judge the effect of these. Pro Tools 11 provides various new commands that allow you to selectively bypass Inserts in your session.

From the Track menu, choose Bypass Insert, then select All to bypass all the inserts in your session; select Inserts A–E to bypass Inserts A–E on all tracks in the session; or select Inserts F–J to bypass Inserts A–E on all tracks in the session.

If you want to bypass specific types of plug-ins, you can choose Reverb, Delay, or Modulation to bypass these categories of plug-in on all tracks in the session.

You can also bypass all EQ or Dynamics plug-ins in the session by choosing these categories. There are some plug-ins, such as Channel Strip, that fall into both of these categories. If this is the case, choosing EQ only bypasses the EQ portion of the plug-in and choosing Dynamics only bypasses the Dynamics portion of the plug-in.

If you prefer to use keyboard commands rather than menu commands, you can Control-click (Mac) or Start-click (Windows) an Insert Assignment button in the Mix or Edit window to bypass this insert and all inserts below it on a track.

If you want to bypass all the plug-ins in your session with a single click, hold the Option (Mac) or Alt (Windows) key while you use the above command. Alternatively, you can press Shift + A to bypass all inserts.

To bypass all inserts A–E, press Shift + 1 or to bypass all inserts F–J, press Shift + 2. To bypass all EQ plug-ins, press Shift + E; for dynamics plug-ins, use Shift + C; for Reverb plug-ins press Shift + V; for Delay plug-ins use Shift + D; and for Modulation plug-ins, press Shift + M.

> **NOTE**
> To un-bypass inserts, simply apply the Bypass commands again.

Add New Track Enhancements

Pro Tools 11 lets you add new tracks simply by double-clicking in the Mix or Edit windows.

To add a new track of the same type and channel width as the last new track: double-click on the empty area of the Edit window below any current tracks; or on the empty area of the Mix window below or to the right of any current tracks; or in the empty area below any current tracks in the Tracks list. If no tracks exist in the session, a stereo audio track is created by default.

To ensure that an audio track is created (rather than another type of track if the last new track was not an audio track), hold the Command (Mac) or Control (Windows) key while you double-click.

If you want to add an Auxiliary Input track of the same channel width as the last new track, then hold the Control (Mac) or Start (Windows) key while you double-click; to add a stereo instrument track, hold the Option (Mac) or Alt (Wndows) key while you double-click; and to add a Master Fader, hold the Shift key while you double-click (this only works on the Mac).

Enable Automation in Record

Pro Tools 11 has a new 'Enable Automation in Record' option in the Operation preferences – see Figure 8.2. This lets you write Automation while you are recording, unlike previous versions that would play back existing Automation while recording, but would not let you create new Automation moves while recording. When this option is enabled, Pro Tools works with Automation during recording exactly as it does when in playback.

Figure 8.2
Enabling automation
in record in the
operation preferences.

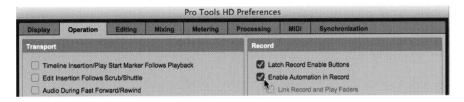

> ### NOTE
> When the 'Enable Automation in Record' option is selected, the 'Link Record and Play Faders' option is automatically disabled and is unavailable.

Transport Fade-in

Yet another useful new feature for Pro Tools 11 is the Fade-in option that has been added to the Transport – see Figure 8.3. When this is highlighted, all the audio in the session automatically fades in, linearly, from 0 dB up to the current levels of the source audio, from when you hit Play. Using the numeric field to the right of the Fade-in enable button. you can set the Fade-in time using to anywhere from zero to four seconds.

This Fade-in feature provides an easy way of avoiding pops and clicks at the start – or sudden boosts in volume when relocating the playback location on-the-fly during playback. For most music projects, you would set a short fade-in of, say, 250 ms. For post-production projects, especially when synchronizing to external equipment that may take a couple of seconds to lock up, you could set longer times.

Figure 8.3
Transport Fade-in.

> **NOTE**
> Be sure to disable the Fade-in Transport option when recording. Audio that is recorded with the Fade-in Transport option enabled will include a fade based on the set Fade-in duration.

Click II

One of the humblest, yet crucially important features of any digital audio workstation is an audible click that can be used during session playback to serve as a tempo reference when performing and recording.

At last, the original Click plug-in has been revamped as a Native, mono-only, AAX plug-in – Click II – with a much larger range of click sounds to choose from – see Figure 8.4. Now you can select a wood block, a shaker, or even the classic UREI click that has been a standard in the film industry for many years.

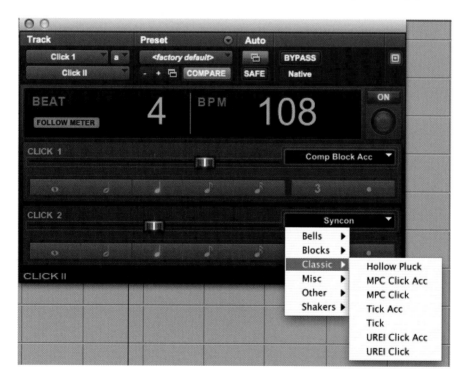

Figure 8.4
Click II showing menu of click sounds.

The Click II plug-in receives its tempo and meter data from the Pro Tools application, letting it follow any changes in tempo and meter in a session.

If the 'Follow Meter' button is enabled, the clicks follow whatever meter has been set – 4/4, 3/4, 6/8, or whatever. If this is disabled, you can manually choose the click settings directly in the plug-in by clicking on the appropriate beat sub-divisions for click 1 (the accented first click of the bar) and for click 2 (the unaccented subsequent beats in each bar). Additional buttons let you choose triplet or dotted values for these clicks.

The plug-in also has a visual indicator at the top right of the window that flashes in time with the click when it is switched on.

The Workspace

The Browser

The Workspace is a file 'browser' – a type of database where you can search through all your samples, loops, and other media files on connected storage drives and audition these before bringing them into your session. The database functions have been significantly improved so that it starts searching as soon as you start typing and delivers the results and relinks files much more quickly than with previous versions.

With previous versions of Pro Tools, there were three types of browser: the Catalog browser, the Project browser, and the main Workspace browser. Pro Tools 11 combines the functions of all these browsers into a single Workspace – see Figure 8.5. You can have several of these browsers open simultaneously if

Figure 8.5
The workspace
browser.

you like. So, for example, you could organize these to display the Session files in one, a folder containing sampled loops in another, and a catalog containing sound effects in a third.

Using the Workspace Browser

You can open the Workspace browser from the Window menu, or by holding the Option key and pressing the semi-colon key (;) on Mac, or by holding the Alt key and pressing the semi-colon key (;) on Windows. To open a second (or subsequent) Workspace browser, Command-double-click (Mac) or Control-double-click (Windows) a volume, catalog, or folder in the currently open Workspace browser.

The Workspace browser window is divided into two sections – the Locations 'pane' at the left and the Workspace 'pane' to the right of this. In the Workspace browser, you can use the Locations pane to look through the 'volumes', that is, the local and networked drives, on your system, or to look through the indexed Catalogs of media or the system User directory. If you are looking for folders or files on your Desktop, for instance, you can find these in the User directory's Desktop folder.

The Workspace pane functions much like DigiBase browsers from previous versions of Pro Tools, but be aware that the Browser menu (or Right-click menu) options vary depending on the selected item. For example, if you select any folder or file in the Workspace and click on the Browser menu, or simply Right-click the item, you can access commands such as Reveal in Finder (Mac) or Reveal in Explorer (Windows), or Move to Trash (Mac) or Move to Recycle Bin (Windows).

> **NOTE**
> When any of these commands are applied to folders, they apply to all sub-directories and media files within that folder.

Similarly, if you select any volume in the Locations pane and click on the Browser menu, or simply Right-click the item, you can access commands such as Delete Index for All Offline Volumes, or Show Offline Volumes.

NOTE
When any of these commands are applied to a volume, they apply to all directories and files on that volume.

And if you select any Catalog in the Locations pane and click on the Browser menu, or simply Right-click the item, you can access commands such as New Catalog or Loop Preview.

NOTE
When any of these commands are applied to a catalog, they apply to all files in that catalog.

Creating a Catalog

To create a Catalog, select a number of folders or files in the Workspace pane and choose 'Create Catalog from Selection' from the Workspace mini-menu located at the top right of the Workspace browser window – see Figure 8.6.

Figure 8.6
Creating a catalog from a selection of audio files.

A dialog box will appear that allows you to name your new catalog – see Figure 8.7

Figure 8.7
'Name the catalog' dialog.

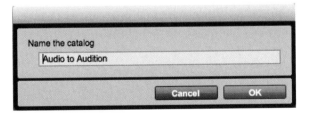

When you OK this, you will see a new folder listed under Catalogs in the Locations pane, and if you open this folder, you will see the files or folders that you have chosen to be in your new catalog – see Figure 8.8 for an example of how this might look.

Figure 8.8
A newly created catalog.

Presets

When you have chosen what information to display in your Workspace Browser, you can save this view as a preset by Command-clicking (Mac) or Control-clicking (Windows) on any of the five preset buttons at the top left of the window.

Unmount Volumes

Pro Tools 11 lets you use the Finder (Mac) or Windows Explorer (Windows) to mount or unmount volumes while Pro Tools is running. Consequently, the Unmount command that was present in lower versions of Pro Tools has been removed.

Database File Management

It is important to know where Pro Tools stores the database information, so that you can make backups of this information to use in the event that you need to re-install these files.

These database files can become corrupted on a hard drive, resulting in slower or erratic behaviour. You can always delete a corrupted database file then relaunch Pro Tools, in which case a new database file will be regenerated. However, it does take some time to build the database files, especially if you have several volumes connected to your system.

TIP

If you have a recent backup of an uncorrupted database file, this can be used to quickly replace a corrupted file – getting your system fully back into operation with a minimum of 'downtime'.

Pro Tools 11 uses a single database file, Workspace.db, that maintains an index of all volumes (hard drives, networked drives, and removable media) and catalogs.

There is also a global waveform cache file for audio files indexed by the database – WaveCache.wfm. This is used to store audio file metadata when no session is open, or for files that reside on read-only volumes. This includes the mini thumbnail waveforms used for previewing. As in previous versions of Pro Tools, this continues to be a separate file that is stored next to the database file – see Figure 8.9.

On the Mac, database files are stored in the following directory: /Users/Shared/Pro Tools, whereas on Windows, database files are stored in: <Systemdrive>/Users/Public/Pro Tools. Individual waveform cache files for sessions continue to be stored in each session folder.

Figure 8.9
Pro Tools database
files for Mac OSX.

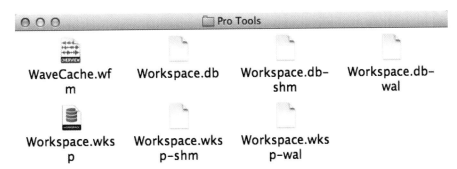

NOTE

With Pro Tools 11, the database file is versioned and is not backwards compatible with previous versions of Pro Tools. When using a system where Pro Tools 11 and Pro Tools 10 are co-installed, there will be separate database files for each version of Pro Tools.

TIP

Database files of the same version can be used cross-platform – on Mac or Windows systems – which can be useful if you need to replace a Mac by a Windows computer, or vice versa, but wish to avoid having to rebuild the database file.

Indexing

Pro Tools automatically indexes (or updates the index for) volumes and directories in the background while browsing. The index for any given session folder (as well other audio/video files folders in other locations) is updated when you open that session.

To manually index a volume or directory (folder), simply Right-click the volume or folder in the Locations pane and choose Updated Database for Selected.

Deleting Volumes and Catalogs from the Database

Pro Tools 11 lets you delete the database index of a volume or catalog, whether or not it is online, that is mounted.

To delete the index for any volume or catalog, Right-click the volume or catalog in the Locations pane and choose Delete Index for Selected.

To delete the index for an offline volume, Right-click anywhere in the Locations pane, select Show Offline Volumes, then Right-click an offline volume and choose Delete Index for Selected.

To delete the indexes for all offline volumes, Right-click any volume in the Locations pane and choose Delete Index for All Offline Volumes.

Enhanced Database Searching

Pro Tools 11 provides an updated user-interface and an enhanced search engine for the Workspace browsers that allows you to search your entire system, or specific volumes or directories, with ease.

Using Simple Search

A simple text search using the Browser Search Field lets you quickly search all indexed volumes for text across all text in the browser – see Figure 8.10.

Figure 8.10
A Simple Search – type name into search field.

Most of the time, you will only want to search for media files, so click the Simple Search button and choose the Search for Media Files Only option in the Browser menu. Then use the Locations pane to select the Volume, Catalog, Session, or Folder that you wish to search – or don't make any selection if you wish to search the entire system, including any offline volumes that are shown.

As soon as you type the names of files, dates, durations, and so forth that you wish to search for into the Browser Search field, Pro Tools starts searching for matches to the text you type and results will start appearing in the Browser pane. How fast the search is will, of course, depend on how much data have to be searched through and how fast your computer system is.

Advanced Search

Pro Tools 11 also provides advanced tools for searching files and volumes by specific categories.

NOTE
When either a simple search or an advanced search is initiated, the Search button changes from a magnifying glass icon to an 'X'. Click the Search button again to end the search.

To initiate an advanced search, either click the Advanced Search button or, if you prefer to use a keyboard command, press Command + F (Mac) or Control + F (Windows).

In the Locations pane, select the Volume, Catalog, Session, or Folder that you wish to search, select the Column type and the Search Criteria, then click in the Search Text field, and type the text for your search. When you press Enter, the search results will be displayed in the Browser pane – see Figure 8.11 for an example of how this might look.

If you wish to refine your search, click the Add Row button to add further constraints to the search and select these before entering your search text.

To switch from an advanced search back to a simple search, simply click the Advanced Search button once more or, if you prefer to use a keyboard command, press Command + F (Mac) or Control + F (Windows) again. To cancel the search entirely, just click the Simple Search button (the 'X').

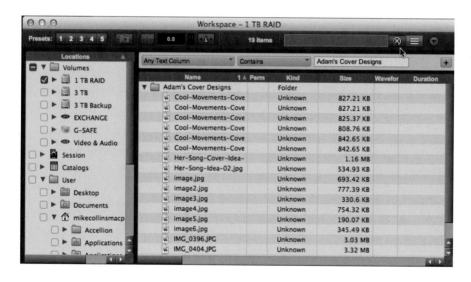

Figure 8.11 Advanced search for Adam's Cover Designs.

Launching and Using Pro Tools 11 for Mac OSX

Network Access Alert When Launching Pro Tools for Mac

If the Firewall in the Security & Privacy System Preferences is enabled, Mac OS X posts a message asking if Pro Tools should allow incoming network connections every time you launch Pro Tools – see Figure 8.12.

Figure 8.12
Network access alert.

Figure 8.12
Network access alert.

According to Avid's Simon Sherbourne, 'This is now standard MacOS behaviour when an application that can make network connections is launched for the first time. It's because PT has built-in network functions in the Marketplace menu and also has Gobbler integration'.

Pro Tools Menus and Interface are Grayed Out on Launch

Pro Tools can appear unresponsive at launch, with the menus grayed out and browsers inoperable. This can happen because of Spaces. If Pro Tools is launched in any other Space than the first one, the Session Quick Start dialog still opens in the first Space. You can temporarily disable Spaces, launch Pro Tools, and disable the Session Quick Start dialog in the Preferences to avoid this problem.

> **NOTE**
> If you manually add Pro Tools to the 'allowed incoming connection' list in Firewall Options, the message may not appear on launch, but in some cases, this setting does not persist. This scenario can be avoided by turning off the Firewall.

Pro Tools Cannot Record to Drives Formatted as Case-Sensitive

Pro Tools cannot record to Mac OS X drives that have been formatted as 'Case-Sensitive'. Format the Mac OS X record volumes as 'Mac OS X Extended (Journaled)' in order to record properly.

Enable Spotlight for Best Performance When Relinking Files and Indexing in the Workspace

For best performance with relinking and Workspace indexing, enable Spotlight. While enabling Spotlight is not required, it will improve performance in this area with Pro Tools.

Mac OS X 10.7.x – 10.8.x Keyboard Shortcuts for Mission Control Conflict with Pro Tools Keyboard Shortcuts

By default, Mac OS X 10.7.x and 10.8.x use Control + Arrow keys to open Mission Control and Application Windows. These keyboard shortcuts conflict with the Pro Tools keyboard shortcuts for changing track display height. You can disable or reassign these shortcuts in Mac OS X.

To disable or reassign keyboard shortcuts in Mac OS X: From the Apple menu, choose System Preferences. Click Keyboard. Click the Keyboard Shortcuts tab. Disable or reassign the shortcut for Mission Control.

In Mac OS X 10.7.x – 10.8.x, the Library Folder in Each User Folder is Hidden

DigiTrace logs, which can be helpful in troubleshooting Pro Tools errors, are stored in User/Library. These logs will be more difficult to find if the Library folder is hidden.

To find the Library folder, go to the Finder and click the Go menu while holding the Option key. Library will appear in the menu while the Option key is held, allowing you to select this.

To display all hidden files and folders in an Open dialog: Press the Command and Shift keys together.

To permanently unhide the Library folder, launch the Terminal application that can be found inside the Utilities folder located within the Applications folder inside the Mac's boot drive. When the Terminal command-line interface appears, type the following:chflags nohidden/Users/<Username>/Library.

Known Issues with Pro Tools 11 for Mac OSX

Limitations of the Avid Coreaudio Driver with CoreAudio Manager

The 003 and Digi 002 systems have the following limitations:

Pro Tools HD and Pro Tools with 003 and Digi 002 systems require exclusive access to Pro Tools hardware. You cannot use CoreAudio applications and Pro Tools with these systems at the same time. To use Pro Tools, make sure you quit any CoreAudio applications before launching Pro Tools on these systems. To use a CoreAudio application with these systems, make sure you quit Pro Tools before starting the CoreAudio application.

The Digi CoreAudio Driver cannot be used to preview sound files from the Mac Finder while Pro Tools is running. When a sound file is located in the Mac OS X navigation window, a QuickTime transport bar is displayed next to it. The QuickTime transport bar lets you audition the sound file. With USB audio interfaces, you can preview sound files from the Mac Finder as long as Pro Tools is not running.

Regardless, with FireWire and Avid HD audio interfaces, the sound will always play back though the Mac built-in audio controller (through the Mac speaker or headphone jack). However, if you double-click a sound file, the QuickTime application will launch, which can use the Digi CoreAudio Driver for playback.

The CoreAudio Driver cannot be used for playback of Mac System Sounds.

Selecting 'Digidesign HW' for Sound Input or Output in System Preferences May Cause Pro Tools to Fail to Launch

If 'Digidesign HW' is selected as the input or output device in System Preferences, certain common actions in the Finder may activate the CoreAudio Manager. These actions include selecting an audio or video file while in Columns view or while the Inspector window is open, or using Get Info or Quick Look on an audio or video file. Once activated, CoreAudio may remain connected even after selecting a different file or closing all Finder windows. If Pro Tools is launched while CoreAudio is connected, an error message stating 'Hardware is in use by another application. (-1133)' may occur.

Pro Tools Hardware May Not Play Sound from Quicktime or Safari with CoreAudio

If you do not hear sound when playing back audio through Pro Tools hardware from QuickTime or Safari with the Digi CoreAudio Manager, try one of the following:

- Open the Mac System Preferences and click Sound. While QuickTime or Safari is playing back audio, change the Output from Digidesign HW to Internal, and then back to Digidesign HW.

- Launch iTunes before launching QuickTime or Safari.

Reconnecting a 003 or 002 After Losing the Connection with Pro Tools

If your 003 family or 002 series interface loses its connection with Pro Tools, increasing your buffer size may bring the device back online without having to close Pro Tools or restart your computer.

Renaming Pro Tools Aggregate I/O Driver in AMS Corrupts the Driver

If you rename the Pro Tools Aggregate I/O driver in Audio MIDI Setup (AMS), the driver may be corrupted the next time you launch Pro Tools. If this happens, quit Pro Tools and delete the Pro Tools Aggregate I/O device from AMS. Pro Tools will re-create the driver the next time it is launched.

Pro Tools Quits Unexpectedly When Enabling the Built-in Digital I/O

When enabling built-in digital I/O for Pro Tools Aggregate I/O in Audio MIDI Setup, Pro Tools quits unexpectedly. To avoid this problem, enable the built-in digital I/O in Audio MIDI Setup before launching Pro Tools.

Index